W9-AZE-871

Jim BABCOCK

The X Series
Recommendations

McGraw-Hill Series-Computer Communications

In order to receive additional information on these or any other McGraw-Hill titles, in the United States please call 1-800-822-8158. In other countries, contact your local McGraw-Hill representative.

The X Series Recommendations

**Standards for
Data Communications**

Second Edition

Uyless D. Black

McGraw-Hill, Inc.

New York San Francisco Washington, D.C. Auckland Bogotá
Caracas Lisbon London Madrid Mexico City Milan
Montreal New Delhi San Juan Singapore
Sydney Tokyo Toronto

Library of Congress Cataloging-in-Publication Data

Black, Uyless D.
    The X series recommendations : standards for data communications /
Uyless Black—2nd ed.
        p.   cm.
    Includes index.
    ISBN 0-07-005593-9 (H)
    1. Computer network protocols—Standards.   2. Data transmission
systems—Standards.   3. Computer interfaces—Standards.   I. Title.
TK5105.5.B5675 1995
004.6'2—dc20   94-40177
                                                              CIP

Copyright © 1995 by McGraw-Hill, Inc. Printed in the United States of America.
Except as permitted under the United States Copyright Act of 1976, no part
of this publication may be reproduced or distributed in any form or by any
means, or stored in a data base or retrieval system, without the prior written
permission of the publisher.

1 2 3 4 5 6 7 8 9   DOC/DOC   9 9 8 7 6 5

ISBN 0-07-005593-9

*The sponsoring editor of this book was Brad Schepp, the manuscript ed-
itor was Aaron Bittner, and the executive editor was Robert Ostrander.
The production supervisor was Katherine Brown. This book was set in
ITC Century Light. It was composed in Blue Ridge Summit, Pa.*

Printed and bound by R. R. Donnelley & Sons Company, Crawfordsville, Indiana.

*Product or brand names used in this book may be trade names or trademarks.
Where we believe that there may be proprietary claims to such trade names or
trademarks, the name has been used with an initial capital or it has been capi-
talized in the style used by the name claimant. Regarless of the capitalization
used, all such names have been used in an editorial manner without any intent
to convey endorsement of or other affiliation with the name claimant. Neither the
author nor the publisher intends to express any judgment as to the validity or le-
gal status of any such proprietary claims.*

Information contained in this work has been obtained by McGraw-
Hill, Inc. from sources believed to be reliable. However, neither
McGraw-Hill nor its authors guarantee the accuracy or
completeness of any information published herein and neither
McGraw-Hill nor its authors shall be responsible for any errors,
omissions, or damages arising out of use of this information. This
work is published with the understanding that McGraw-Hill and
its authors are supplying information but are not attempting to
render engineering or other professional services. If such services
are required, the assistance of an appropriate professional should
be sought.

MH95
0055939

*To my friends Joe and Marie Santamauro
and Joe, Mike, and Trina.*

Contents

Preface

This book examines the International Telecommunication Union-Telecommunication Standardization Sector or ITU-T (formerly known as International Telegraph and Telephone Consultative Committee or CCITT) Recommendations pertaining to the X Series. These recommendations (also called standards by many people) have become some of the most widely used specifications in the world for defining how data are exchanged between computers. In the past, the X Series Recommendations were used principally in Europe because of the influence of the Postal, Telephone, and Telegraph (PTT) administrations in each European country. However, with the growing recognition of the need for international communications standards, the X Series have found their way into practically all countries of the world and into practically all vendors' products. Their use has paved the way for easier, more efficient, and less costly communications between computers, terminals, and other data processing machines.

The X Series Recommendations contain over 3000 pages of detailed (and changing) specifications. Obviously, the volume, complexity, and dynamic nature of the Series presents problems in staying abreast with the material. Of course, many of the X Series Recommendations do not pertain to a person's job or perhaps they are simply of no interest. Notwithstanding, practically any data communications professional must come to grips with many of these standards and therefore the sheer amount of text can be somewhat overwhelming.

This book will aid the reader in learning about these important standards. The book is designed to provide the reader with a tutorial on each of the X Series Recommendations. It is also designed to provide a reference guide to each of the X Series Recommendations.

The most difficult task I faced in writing this book was choosing which material to include from the many pages that comprise the X Series. The decision was based on concentrating on those recommendations that are more widely used throughout the world and the newer recommendations that are gaining support and experiencing increased use. Notwithstanding, each X Series Recommendation is described in this book, albeit some are examined in more detail than others.

Since this book is meant as a tutorial and a general reference guide on the ITU-T X Series Recommendations, you cannot design nor can you implement an X Series-based system after only reading this book. It should not be used as a replacement for reading and studying the ITU-T X Series Recommendations. Notwithstanding, it is hoped that this material will make the X Series documents more understandable to the reader and make the reading of the X Series material a little less painful.

"Notes for the reader"

The International Telegraph and Telephone Consultative Committee (CCITT) has been reorganized and renamed (Appendix 1A in chapter 1 provides more details). It is now called the International Telecommunication Union-Telecommunication Standardization Sector (or ITU-T).

Periodically, the ITU-T declares out-dated Recommendations "no longer in force" (NLIF), which means they are not supported by the ITU-T, and the ITU-T recommends they be discontinued in vendor products. Notwithstanding, they may still be used in some vendor products, and may be important to the reader. Therefore, this book includes the more recent NLIF Recommendations. In the chapter in which they are described, they are identified with the initials NLIF for: no longer in force.

The V Series is undergoing revisions, deletions and additions continuously. For those Recommendations that have not been completed (as of this writing), they will be noted in the appropriate chapter with UCFA, which are the initials for: under consideration for approval.

The initials of NLIF and UCFA are created for this book, and are not used by the ITU-T.

Uyless Black

Acknowledgments

I would like to thank Holly Waters for her expert assistance in preparing this book. On numerous occasions, she made suggestions that improved the text. She also read the material several times and corrected inconsistencies in several of the sections. Her dedication made this a better book.

A note of thanks is due to the ITU-T for their ongoing efforts in promoting and publishing standards for data communication systems and networks. These standards can only help companies and organizations in their efforts to increase productivity and reduce the complexity of their communications interfaces and overhead. This book has been prepared independently of ITU-T and reflects the views of the author and not necessarily of ITU-T.

The ITU-T is the copyright holder for the X Series Recommendations. The excerpting and reproduction of any material is authorized by the ITU-T organization, the copyright holders. The choice of any excerpts is that of the author and does not affect the ITU-T in any way. The full text of these recommendations can be obtained from the ITU Sales Section, Place des Nations, Geneva, Switzerland.

1

Introduction to the X Series

The goal in this chapter is to discuss the rationale for the use of standards in the data communications industry, to define some of the pros and cons of their use, and to set the stage for chapter 2, which introduces the X Series Recommendations. The reader who is conversant with this topic may proceed directly to chapter 2.

The Growing Use of Standards

Computer communications systems are used to support the transfer of data between two *end users*. An end user may be a person or even a computer or terminal. Because the transfer is data (instead of voice or other images), the process is called data communications (some organizations prefer the term to be singular, as in data communication).

In order for this seemingly simple process to occur, the computers and their communications facilities must perform many actions. For example, the two machines must accept each other's file formats (in what form the data appears in the file), and they must agree on the syntax of the data (text, numeric, the number of decimal points, etc.). If an intervening network is supporting the session between the machines, it must know about their communications characteristics (that is, the procedures each machine uses to connect to the network, such as passwords, requests for services, etc.). If the data are important, some means must be provided to assure that the data transfer through the network is successful and that both machines know that all data has been transferred without any problems.

These functions are not trivial, and they require a considerable amount of software and hardware to support them. Yet computers usually communicate without ambiguity if (1) they are instructed (programmed) correctly, (2) the communications signals between them are not distorted, and (3) the computers understand the intent and meaning of each other's symbols. This last requirement presents a formidable challenge because it implies a high degree of understanding and cooperation between the machines. In effect, it implies that the computers communicate with a common set of symbols and, equally important, an unambiguous interpretation of these symbols. It implies the use of *standards*.

Standards not only ease the task of interfacing different computers, they also give the user more flexibility in equipment and software selection. In addition, the acceptance and use of a standard often leads to lower costs to consumers, because the standard can be written into and sold with off-the-shelf software and even implemented with very large-scale integrated (VLSI) chips.

The Origin of International Data Communications Standards

The ITU-T has been involved in the development and promulgation of communications standards for many years. In the earlier part of this century, the focus for international telecommunications standards was on the telegraph, and later the telephone. The ITU-T assumed the role as the authoritative body for promulgation of standards on these two technologies. As the use of computers and computer-based networks increased in the 1970s and 1980s, the ITU-T began the development of other standards in this important area. The X Series is one of these standards.

The X Series Recommendations owe their origin to the recognition of the need for compatible communications architectures between different manufacturers' data communications protocols.

The early computers that provided communications services were relatively simple. In a typical configuration, terminals were connected to a computer in which several software programs controlled the communications process by transmitting and receiving data via a telephone line. The line was usually attached to an interface unit within or connected to the computer. The unit provided signalling (how the symbols were coded) and synchronization (how the symbols appeared on the line).

These early systems used conventions based on the telegraph and telex applications and transmitted messages with special codes, such as the so-called Baudot five-digit code. Later, a 7-bit character code was developed for teletype machines. This code is known as International Alphabet Number 5 (IA5), and is also associated with the ASCII code. These codes were often used and interpreted differently by the manufacturers of communications products. Some companies, such as IBM, introduced their own codes into their product line. One of these codes, EBCDIC, has also become a *de facto* industry standard.

These early activities led to compatibility problems between the different vendors' terminals and computers. If two machines were to communicate with each other, they had to use the exact same code set. If they did not have the same code, some form of conversion process had to take place in one or both of the computers and/or terminals. The problem was quite similar to a situation in which two humans attempt to communicate with each other while using different languages. Worse yet, each vendor began the development of proprietary protocols that were incompatible with those of other vendors.

The end user and customer began to question this environment. As a result, user groups sprouted up across countries. Their purpose was to provide input into a vendor's decisions about products. The end user also began to question the necessity for vendor-specific closed systems. After all, other industries had a baseline set of standards (building codes for houses, standards for highway construction, etc.), why not the computer and communications industry as well?

The ITU-T assumed the lead role in defining the procedures for using the telephone system for transmitting data, with the publication of the V Series Recommendations. Shortly thereafter, it began the effort to develop the X Series.

Other organizations such as the ISO, IEEE, ANSI, and EIA also began working on the development of standards. Many of these efforts have found their way into the X Series Recommendations that are examined in this book.

Commonly Used Terms in the X Series Recommendations

This section introduces and explains some terms that are used in the X Series Recommendations. They are rather basic, but the ITU-T documents assume the reader understands them.

The term *network* in this book is used to describe the organized connection of machines for the purpose of exchanging data. The term *network architecture* is commonly used today to describe networks. Paraphrasing the dictionary definition, an architecture is a formation of a structure. A network architecture describes the hardware and software in the network and the interfaces with the network. An architecture also encompasses data link controls, standards, topologies, and protocols.

Like architecture, the term *protocol* is borrowed from other disciplines and professions. A protocol defines conventions and rules about how network components establish communications, exchange data, and terminate communications.

The term *topology* is used to describe the form of a network. A network topology is the shape (or the physical connectivity) of the network.

Figure 1.1 shows the relationship of the *user access line* (UAL) to the networks. In this figure, user A connects its data terminal equipment (DTE, which is an end user computer) to a packet-switched network through the UAL and a *packet switch* (or a similar device, such as a circuit switch). ITU-T uses the term *data circuit-terminating equipment* (DCE) to describe the DTE interface with the network. Typically, this DCE is a modem, a multiplexer, or some type of digital service unit. From the perspective of ITU-T, the network interface extends to the user DCE.

This figure also illustrates the *internetworking unit* (IWU). It is used to interconnect networks. The network on the left of the figure is a packet data network. It uses conventional switching technology to route traffic through the network. The network on the right is an example of a broadcast network (in this figure, it is a local area network). In this example, the traffic is sent to all the devices attached to the network (DTEs B, C, D, and E). In turn, these devices examine the address in the protocol data unit (PDU). If it is destined for the device, it copies and passes this packet to an upper layer protocol. If not, it simply ignores the packet.

Figure 1.2 is a more detailed view of the user access line, which connects to the packet exchange and the user device (DTE) through modems, digital service units, multiplexers, etc. The ITU-T uses the term *DCE* to define the device at the user premises furnished by the Postal, Telephone and Telegraph Administration (PTT) (or other network supplier). From this perspective, the user only "sees" the operations at the local interchange circuits (ICs), identi-

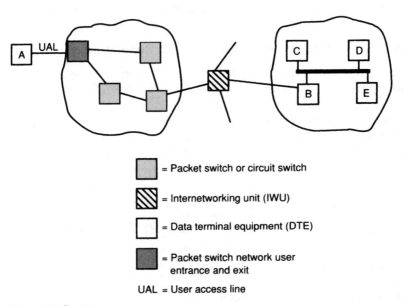

    ☐ = Packet switch or circuit switch

    ▨ = Internetworking unit (IWU)

    ☐ = Data terminal equipment (DTE)

    ■ = Packet switch network user entrance and exit

UAL = User access line

Figure 1.1 Basic terms.

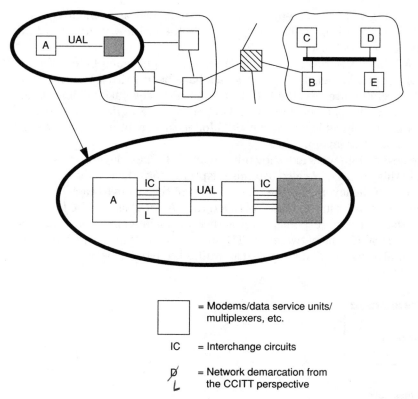

Figure 1.2 User access line.

fied in the figure with the uppercase L. This approach works well enough in countries where the PTT furnishes the communications line, modems, and all other components that extend to the user's machine. However, in some countries, this setup consists of components offered by more than one supplier. Indeed, in a private network, the end user must be aware of the operations inside the network. The point of this discussion is that the reader should check with the network supplier to determine where the demarcation of the network begins and ends. In some countries, the demarcation is dictated by the PTT.

While the terms DTE and DCE are useful to depict a conceptual operation, in many actual situations the functions of the DTE and DCE are housed in one machine. For example, a large-scale mainframe computer or front-end processor may contain the functions of DTEs and DCEs on the same line card.

The X Series and the Open Systems Interconnection (OSI) Model

The ITU-T X Series Recommendations have played a major role in fostering common data communications standards among different vendors and manufacturers. The recommendations (also known as standards) have been accepted in Europe for many years. In the last few years, they have achieved worldwide use.

Many of the X Series Recommendations were published before the advent of the OSI Model. The recent series have been written to conform to the OSI architecture, and the ITU-T has published several guidelines to aid in using the "old" specifications with OSI. Since the X Series Recommendations are found in all layers of the OSI Model, this chapter provides a brief review of the OSI layers.

The seven OSI layers are shown in Figure 1.3. These layers are summarized in this chapter. We also use this chapter to introduce some important attributes of the layers and the relationships of the layers to each other.

The lowest layer in the model is called the *physical* layer. The functions within this layer are responsible for activating, maintaining, and deactivating a physical circuit between a DTE and a modem, multiplexer, or some other similar device. The layer also identifies the bits (as 0s or 1s).

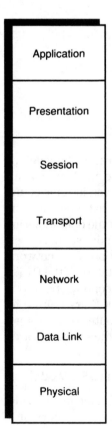

Figure 1.3 The OSI Model.

This layer is concerned with the nature of the signals. Consequently, it must be able to distinguish between different levels of voltages and the direction and intensities of currents. This layer has the task of sending and receiving electromagnetic signals, as well as creating and interpreting the optical signals in optical fibers. It is also responsible for defining the cabling and wiring between machines (if any exist). This layer also contains the specifications for the physical connectors used to attach cables and wires to the computers (connectors that some people simply call plugs).

The *data link* layer is responsible for the transfer of data over the channel. It provides for the synchronization of data to delimit the flow of bits from the physical layer. It also provides for the identity of the bits within transmissions. It ensures that the data arrives safely at the receiving computer or terminal. It provides for flow control to ensure that the computer does not become overburdened with too much data at any one time. One of its most important functions is to provide for the detection of transmission errors and to provide mechanisms to recover from lost, duplicated, or damaged data.

The *network* layer specifies the interface of the user DTE into a data communications network, as well as the interface of the DTEs with each other through a network. The network layer is the layer responsible for routing and route discovery. In a packet switch, this routing function is called packet switching; in a circuit switch network it is called circuit switching; frame relay networks use the term frame relay switch; and cell relay networks use the term cell relay switch.

The network layer is responsible for supporting the end user's negotiation of services with a network. These services are referred to as quality of service (QOS) features. They include functions such as the reverse-charging of calls (like the collect call service in a telephone system), negotiating the size of the data unit traversing the network, providing for security features, and so on.

The *transport* layer provides the interface between the data communications network and the upper three layers. This layer gives the user options in obtaining certain levels of quality (and cost) from the network itself (i.e., the network layer). It is designed to keep the user isolated from some of the physical and functional aspects of the packet network. It also provides for end-to-end integrity of the transfer of user data.

The *session* layer serves as the user interface into the transport layer. The layer provides for an organized means to exchange data between end user applications. The users can select the type of synchronization and control needed from the layer. To cite some examples: (1) Users can establish an alternate two-way dialogue or a simultaneous two-way dialogue, or (2) synchronization points can be established to recover from file and data transfer problems.

The *presentation* layer is used to assure that user applications can communicate with each other, even though they may use different representations for their protocol data units (packets or messages). The layer is concerned with the preservation of the syntax of the data. For example, it can accept various data types (character, Boolean, integer) from the application layer and negotiate an acceptable syntax representation with another peer presentation layer, perhaps located in another computer. It also provides a means to describe data structures in a machine-independent way. It is used to code data from an internal format of a sending machine into a common transfer format and then decode this format to a required representation at the receiving machine.

The *application* layer is concerned with the support of an end user application process. The layer contains service elements to support application processes such as job management, file transfers, electronic mail, and financial data exchanges. The layer also supports the virtual terminal and virtual file concepts. Directory services are obtained through this layer.

Some additional thoughts regarding the seven layers might prove helpful. The lower three layers specify the machine-to-machine communications wherein the machines communicate directly on point-to-point links, multipoint links, or through intermediate systems. Generally, these intermediate systems are networks, whose operations, logically enough, reside in the network layer. Layer four, which provides the "bridge" between the upper three layers and the lower three layers, specifies the end-system-to-end-system communications. The top three layers are concerned with end user communications and direct support of the user applications software.

Connectionless-Mode and Connection-Oriented Communications

The connectionless-mode and connection-oriented services are both widely used operations in data communications networks. Their principal characteristics are:

- *Connection-oriented operations*   Sets up a logical connection before the transfer of data. Usually some type of relationship is maintained between the data units being transferred through the connection.

- *Connectionless-mode operations*   No logical connection is established. The data units are transmitted as independent units.

The connection-oriented service requires a three-way agreement between the two end users and the service provider (for example, the network). It also allows the communicating parties to negotiate certain options and QOS functions. During the connection establishment, all parties store information about each other, such as addresses and QOS features. Once data transfer begins, the user traffic, called protocol data units (PDUs), need not carry all this overhead protocol control information (PCI). All that is needed is an abbreviated identifier to allow the parties to look up the full addresses and QOS features. A connection-oriented system might also perform sequencing, error control, acknowledgments, and flow control on the traffic. The term "might" is emphasized in the previous sentence, because more recent connection-oriented systems, such as frame relay and the asynchronous transfer mode (ATM), perform no acknowledgments of user traffic.

Because the session can be negotiated, the communicating parties need not have prior knowledge of all the characteristics of each other. If a requested service cannot be provided, any of the parties can negotiate the service to a lower level or reject the connection request.

The connectionless-type service manages user PDUs as independent and separate entities. No relationship is maintained between successive data transfers and few records are kept of the ongoing user-to-user communications process through the network(s). The communicating parties must have a prior agreement on how to communicate, and the QOS features must be prearranged or described in each PDU. The more common approach is to carry the QOS parameters in each PDU. The term *datagram* network is also associated with a connectionless network.

Most connectionless protocols do not perform sequencing of the user PDU, nor do they support the positive or negative acknowledgments of traffic at the receiver. Some connectionless systems institute flow control procedures to prevent the network from becoming saturated with traffic, but the most common practice is to discard this traffic, and assume another (at a higher layer) protocol will handle this problem.

By the very nature of connectionless service, it achieves (1) a high degree of user independence from specific protocols within a subnetwork, it achieves (2) considerable independence of the subnetworks from each other, and it achieves (3) a high degree of independence of the subnetwork(s) from the user-specific protocols.

It is intended that most of the layers (as options) support either connection-oriented or connectionless service. (The ITU-T and ISO differ in their approach to this matter.)

Until recently, most of the ITU-T X Series Recommendations were connection-oriented. The connection-oriented approach stems from ITU-T's experience in dealing with connection-oriented telephone systems. Moreover, the PTTs have the task of maintaining strict accountability on the use of their networks, and the connection-oriented approach gives them more tools to use for controlling the use of the network, accounting for the traffic, billing the customer, etc. This is not to say that a connectionless mode of operation could not have software added to it to perform these functions. However, these tasks are more easily accomplished with connection-oriented protocols.

The ITU-T Use of the Registration Hierarchy

The ITU-T and ISO have jointly developed a scheme for naming and uniquely identifying objects, such as standards, member bodies, organizations, protocols—anything that needs an unambiguous identifier. The scheme is a hierarchical tree structure wherein the lower leaves on the tree are subordinate to the leaves above, except, of course, for the root. The example in Figure 1.4 shows the approach used by the ITU-T. The organization uses four arcs below its node to identify its recommendations, questions being studied by study groups, administrations, and network operators. Below these four arcs are other subordinate definitions. The X Series reside at the REC(0) node of the tree. We will use this figure later in this book to explain some registrations for the ITU-T X Series Recommendations.

The Debate on Using Standards

The ITU-T and other standards groups are sometimes criticized because of the lengthy process involved in making a standard available to the public. By the time a standard is available from which to develop a product, the products are already developed by one or more manufacturers because the need to stay abreast of the competition and to capitalize on new technology. Unfortunately, in the vast majority of cases, the different vendors' products are incompatible.

In fairness to the standards groups, it should be recognized that any standard that encompasses a complex technology and tries to satisfy a worldwide group of diverse users necessarily entails a tremendous amount of effort, as well as extensive participation and coordination from many people with different views and interests. This process takes time.

There are two possibilities for the development, publication, and use of international telecommunication standards. First, a standards group could collect proposals from all the diverse groups, analyze them, conduct research and development to determine their feasibility, perform field trials, and then publish the standard based on these activities.

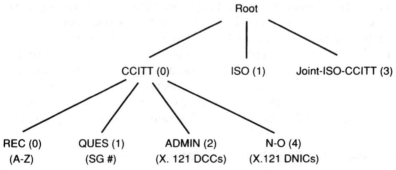

Figure 1.4 The ITU-T registration hierarchy. (The term CCITT, retained until IUT-T publishes specific revision)

The second option is for the standards group to set up a broad framework from which the standard could be developed. The standard is made as detailed as possible, given a time constraint. Consequently, the standard might be issued with certain pieces unfinished (perhaps with the notation "for further study"). The standard is then published. As it is used, the "gaps" are filled in, deficiencies are noted, and modifications are made in the light of practical operating experience. This latter approach is the one usually taken by the ITU-T.

Summary

We have learned that the X Series Recommendations are some of the most widely used data communications standards in the world today. The newer releases of these standards are based on the OSI Model. This chapter provided a very brief summary of the relationship between the X Series Recommendations and the OSI standards.

This chapter also discussed the pros and cons of using standards and summarized the issue by emphasizing that in spite of the slow development process and loss of freedom in implementing the product, the use of standards is a far preferable approach to that of allowing each manufacturer to develop a unique and "closed" system.

With this information in mind, we now examine the structure and organization of the X Series Recommendations. The reader should be aware of the subject matter in this chapter as you read the following chapters. On several occasions, we will refer to some ideas and terms that are introduced in chapter 1.

Appendix 1A: The ITU-T

The ITU-T is a member of the International Telecommunications Union (ITU), a treaty organization formed in 1865. The ITU is now a specialized body within the United Nations. The former CCITT was formalized as part of ITU in 1956. ITU-T sponsors a number of recommendations primarily dealing with data communications networks, telephone switching standards, digital systems, and terminals. Each member country casts a vote on the ITU-T issues.

The ITU-T's Recommendations (also informally known as standards) are very widely used. Until recently, its specifications were republished every 4 years in a series of books that take up considerable space on a book shelf. The 4-year period books are identified by the color of their covers. The colors are used in the following order: red, blue, white, green, orange, and yellow. The 1960 books were red; 1964, blue; 1968, white; 1972, green; 1976, orange; 1980, yellow; and, in 1984, once again red. The 1988 Blue Books "consume" about 3 feet of shelf space.

Changes made at Melbourne

The Melbourne meeting for the Blue Book was a watershed for the ITU-T in that it realized the increasing number of ITU-T standards dictated a change to its organizational structure. First, the Study Groups were restructured (more about this topic shortly).

Perhaps the most significant change that occurred at Melbourne is the administration of the ITU-T Recommendations. The ITU-T Recommendations are now published on a more frequent basis. The conventional 4-year cycle has been eliminated and the standards will be published when a 70 percent or more approval vote is reached by the members. If the proposal is not accepted, it will be sent back to the Study Group.

This approach eliminated the 16,000-page publish-all-at-once mess that has resulted from the previous approach. It also meant the end of the strict delineation of Yellow, Red, Blue Books, etc.

Since the Melbourne meeting, the ITU has reorganized and renamed the CCITT as the International Telecommunication Union-Telecommunication Standardization Sector (ITU-T). In effect, on February 28, 1993, the name "CCITT" no longer existed. The new organization is reflected in Figure 1A-1. The Study Groups (SGs) were also realigned, and their revised names and responsibilities are summarized in Table 1A.1.

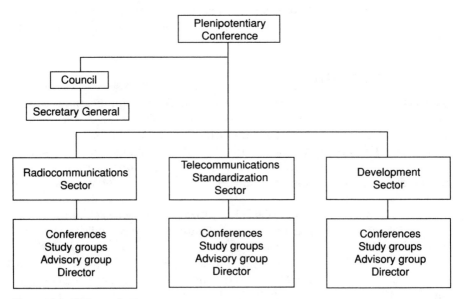

Figure 1A.1 ITU organization.

TABLE 1A.1   ITU-T Study Groups

SG 1	Service definition, operations, and quality of service
SG 2	Operation of network
SG 3	General tariff principles, including accounting
SG 4	Maintenance of all aspect of the network
SG 5	Protection against environmental and electromagnetic effects
SG 6	Outside plant
SG 7	Data communications networks and OSI
SG 8	Terminal equipment for telematic services (fax, teletext, etc.)
SG 9	Television and sound transmissions
SG 10	Languages and methods for telecommunications applications
SG 11	Switching and signalling
SG 12	Transmission performance of networks and terminals
SG 13	General network aspects
SG 14	Transmissions of data over telephone networks (modems)
SG 15	Systems and equipment

The members of the SGs are sent by their respective national telecommunications administrations to participate in the SG activities. In 1993, 181 administrations were represented. However, other participation is allowed, principally by:

- Recognized operating agencies (ROAs), such as AT&T, MCI, and Sprint
- User organizations from industry, such as the airline industry
- Regional standards bodies, such as ETSI in Europe
- Manufacturers of telecommunications equipment
- Scientific and industrial organizations (SIOs)

As stated earlier, the formal documents produced by the ITU-T are called Recommendations, although people are increasingly using the term standards, because many of them do become standards in many countries.

The Study Groups can be very large; sometimes they consist of hundreds of participants. Therefore, SGs are divided into Working Parties, which may be divided further into Expert Groups and Ad Hoc groups. The idea is for these bodies to "do the work" and submit (as much as possible) well-thought-out solutions and documents to the SGs. Each SG has a number of Raporteurs who are responsible for coordinating activity and keeping the process going forward.

Table 1A.2 lists the ITU-T Series and a general description of their scope. The reader may find ITU-T Series A useful. It describes the ITU-T's organization and working procedures.

TABLE 1A.2 ITU-T Recommendations

Series number	Series scope
A	ITU-T organization
B	Definitions, symbols, classifications
C	General telecommunications statistics
D	General tariff principles
E	Telephone operation, network management, and traffic engineering
F	Telegraph, telematic, message handling and directory services, operations, quality of service, and definition of service
G	Transmission systems and media, digital systems and networks
H	Line transmission of nontelephone signals
I	Integrated services digital networks (ISDN)
J	Transmission of sound program and television signals
K	Protection against interference
L	Operations of cable and other elements of outside plant
M	Maintenance: international transmission systems, telephone circuits, telegraphy, facsimile, and leased circuits
N	International sound program and television circuits
O	Specifications of measuring equipment
P	Telephone transmission quality, telephone installations
Q	Telephone switching and signalling
R	Telegraph transmission
S	Telegraph services terminal equipment
T	Terminal equipment and protocols for telematic services
U	Telegraph switching
V	Data communication over the telephone network
X	Data communication networks
Z	Programming languages

Appendix 1.B: Summary of ISO Communications Standards

This appendix provides summary information on the ISO standards. The standards are described by a number and an abbreviated title. The number in the first column is the ISO document number. For purposes of simplicity, the prefixes *ISO, DIS, and DP* have been omitted. The second column is a brief description of the document, which is derived from its title. It is not the title itself.

Be aware that this list is a summary. The documents are grouped around the four-digit identifier, and the titles are a summary of the specific titles of the standard(s). Also, be aware that one identifier may relate to several documents. The list does not include all the ISO standards, because this organization publishes standards on other activities such as financial transactions coding conventions, etc. For a detailed title description, length of document, cost, etc., you should contact the ISO.

Many of the ISO standards are aligned or are closely aligned with the ITU-T Recommendations. However, differences do exist. The differences and alignments are now cited in the 1988 ITU-T Blue Books and subsequent Recommendations (a very welcome service).

OSI Model and OSI management standards

9594

7489   OSI Reference Model, including specifications for security,
        naming and addressing, management, connectionless service
8509   OSI service conventions
9575   OSI routing framework
9594   The Directory use for OSI management
9595   OSI management information service definition
9646   OSI conformance testing
9834   Procedures for registration authorities

Physical layer

9314   (three documents)
2110   The 25-pin connector assignments
2593   The 34-pin connector assignments
4902   The 37- and 9-pin connector assignments
7477   Physical connections using V.24 and X.24 interchange circuits
7480   Start-stop signal transmission quality
8480   DTE-DCE interface backup control
8481   X.24 interchange circuits using DTE provided timing
8482   Twisted pair multipoint interconnections
8877   Interface for ISDN basic access at the S and T reference points
9067   Fault isolation using test loops
9314   Fiber Distributed Data Interface (FDDI)
9543   Synchronous transmission quality at DTE-DCE interface
9578   Communication connectors used in LANs
10022   OSI physical service definition

Data link layer

4335

1745   Basic control mode procedures for data communications
        systems
2111   Basic control mode procedures for code-independent data
        transfer
2628   Basic control mode procedures, complements version
2629   Basic control mode procedures, conversational information
        message transfer
3309   HDLC frame structure and addendum
4335   HDLC elements of procedures
7448   Multilink procedures (MLP)
7776   HDLC-LAPB compatible link control procedures
7809   HDLC-consolidation of classes of procedures; list of standard
        HDLC protocols that use HDLC procedures

8471	HDLC balanced, link address information
8802	Local area network standards, largely derived from the IEEE 802 standards
8885	HDLC-additional specifications describing use of XID frame, and multilink operations
8886.2	Data link service definition for OSI
9234	Industrial asynchronous data link procedures

Network layer

8882	
4731	End system to intermediate system, to be used with 8208
8208	X.25 packet level protocol for the DTE
8348	Network service definition, including addressing conventions, connectionless mode, and additional features
8473	Connectionless-mode network service
8648	Internal organization of network layer
8878	Using X.25 to provide OSI connection-mode network service
8881.3	Using X.25 in local area networks
8882	X.25 conformance testing
9068	Connectionless network service using ISO 8208
9542	Routing exchange protocol to be used with connectionlessmode network service
9574	Operations of a packet mode DTE connected to an ISDN
9577	Protocol identification in the OSI network layer
10028	Relaying functions of a network layer intermediate system

Transport layer

8073	
8072	Transport service definition, connectionless mode
8073	Transport service definition, connection-oriented
8602	Transport service for connection-mode protocol
10025	Conformance testing for connection-mode transport protocol operating on the connection-oriented network service

Session layer

8327	
8326	Session service definitions
8327	Session layer protocols
9548	Connectionless-mode session service

Presentation layer

8823   (two documents)
8822   OSI connection and connectionless presentation services
8823   OSI amendment for PICS *pro forma*
8824   Abstract Syntax Notation 1 (ASN.1)
8825   Basic encoding rules for ASN.1
9576   OSI connectionless protocol to provide connectionless service

Application layer

8650
8211   A data descriptive file for information exchange
8571   File transfer, access, and management (FTAM)
8649   OSI common application service elements (CASE)
8650   OSI protocols for CASE
8831   OSI job transfer and manipulation (JTM)
8832   Basic protocol class for JTM
9007   Concepts and terminology for the conceptual schema and the
        information base
9040   Virtual terminal (VT) protocol
9041   Virtual terminal (VT) protocol
9545   OSI Application layer structure
9804   Application service elements for commitment, concurrency, and
        recovery (CCR)
9805   Protocols for commitment, concurrency, and recovery (CCR)

Appendix 1.C: Overview of Organizations in Standards Promotion and Conformance Testing

The Corporation for Open Systems (COS)

In 1986, several major data processing and data communications suppliers formed a nonprofit venture titled the Corporation for Open Systems to provide a means to accelerate the use of multivendor products that operate under OSI, ISDN, and related international standards. One of its more important activities is the development of a consistent set of testing and certification methods.

The Standards Promotion and Application Group (SPAG)

SPAG is a European-based organization consisting of communications and mainframe vendors. It is concerned with the development and implementation of conformance testing products. SPAG has several conformance testing systems being used in Europe and is now coordinating its activities with COS and POSI.

The Japanese Promoting Conference for OSI (POSI)

The major Japanese computer vendors and the Nippon Telephone and Telegraph Corporation established POSI in 1985 to promote the OSI standards and the efforts of COS and SPAG. POSI is also involved in conformance testing through the Interoperability Technology Association for Information Products (INTAP).

The Australian National Protocol Support Center (NPSC)

The Australian government has been the focal point for the founding of NPSC. As with COS, SPAG, and POSI, this organization is promoting OSI and the development of conformance testing procedures in Australia.

Conformance testing

The European conformance testing is coordinated by the Standards Promotion and Applications Group (SPAG), which has been working for several years on developing conformance testing specifications. The OSI workshop in Europe working under SPAG is the European Workshop for Open Systems (EWOS). In Asia, the SPAG and COS equivalent is Promotion for OSI (POSI) and its workshop is named the Asian-Oceanic Workshop (AOW). These workshop groups (along with the North American group) coordinate their activities and often exchange liaison leaders with respect to conformance testing.

The major conformance testing efforts in the United States are coordinated by the following organizations. The Corporation of Open Systems (COS) and National Institute of Standards and Technology (NIST) are working together in a joint venture to develop conformance testing specifications. These two organizations collaborate with each other to ensure that commercial and U.S. government systems adhere to a test profile. The arrangement is to select abstract test suites for testing by the NIST's National Voluntary Lab Accreditation Program (NVLAP). This program has the task of accrediting the suites. COS has submitted the following suites to NVLAP: FTAM, X.400, Transport protocols classes 0 and 4, X.25, 802.3, 802.4, connectionless network protocol (CLNP).

NIST also sponsors the National Institute of Standards and Technology OSI workshop. It consists of experts throughout government and private industry who work together to further define the OSI documents, to look for gaps in the specifications, and to search out ambiguities in the text.

2

Overview of the X Series

The ITU-T Study Group VII is responsible for the X Series. The X Series are categorized according to the functions and services they provide and are then further classified into the specific X Series Recommendation. Table 2.1 provides a list and a general description of the X Series categories.

The broad categories of services for the X Series are described in the following documents (also see Figure 2.1):

- X.1 - X.39: Services and facilities, and interfaces
- X.50 - X.181: Transmission, signalling, and switching; network aspects; maintenance; and administrative arrangements
- X.200 - X.294: Open Systems Interconnection (OSI) general
- X.300 - X.370: Interworking between networks
- X.400 - X.485: Message handling systems
- X.500 - X.582: Directory services
- X.610 - X.665: OSI networking and system aspects
- X.700 - X.745: OSI management
- X.800 - X.862: Security and OSI Applications

Major Changes in the 1988 Blue Books

Several major changes were made with the publication of the Blue Books. I considered removing this "old" material from this second edition, but some

TABLE 2.1   Organization of the X Series

Category	General description
Services and facilities	Description of recommendations that are used by other X Series and of others that describe various types and categories of service for data communication networks
Interfaces	Recommendations describing physical, data links, and network level interfaces between data terminal equipment (DTE), packet assemblies/disassemblies (PADs), data circuit-terminating equipment (DCE), integrated services digital networks (ISDNs), and data communication networks
Transmission, signalling and switching, network aspects, and maintenance arrangements	Describes a wide range of services, from modulation techniques to multiplexing and control signals
OSI systems	Recommendations that define the service definitions and protocol specifications of the OSI layers
Interworking	Describes the conventions for internetworking various networks, such as packet networks, ISDNs, maritime networks, telephone networks, etc.
Message handling systems	Recommendations that cover the use of message handling systems, sometimes called electronic mail services
The Directory	Several recommendations that define how the directory is to be used to manage an OSI environment
OSI networking and system aspects	Additional information about the relationships OSI, X.25, naming, and ISDN
Security and OSI applications	Security and other OSI applications
OSI management	Jointly published with the ISO, defining the objects in a network

of my clients suggested it should be retained to give the reader a sense of the evolution of the X Series. Upon reflection, I agree. The reader who is interested in the present organization can skip this section.

First, the network interface protocols, which were published in a separate document, have been folded into the document titled, "Network Services and Facilities and Interfaces." Second, the OSI Recommendations were greatly expanded. Third, a new series was published dealing with directory services under the X.500 Recommendations. Fourth, a significant number of the functions embedded in the X.400 Message Handling Systems (MHS) Recommendations were removed from these protocols and placed in the general OSI Model within the application layer.

This decision was made because a considerable number of software modules developed from the MHS protocols can be put to general use. Therefore, it does not make sense to embed them within a specific recommendation. As we shall see later in the book, these protocol specifications and service definitions, defined previously in MHS, are now available as general support services for a number of other applications such as transaction processing systems, database management systems, file transfer systems, virtual terminal systems, etc.

The list below summarizes the major changes made in the Blue Books (relative to the 1984 Red Books):

- The 1984 Red Book used Fascicle VIII.2 and Fascicle VIII.3 to contain X.1X.32. The Blue Book combines these series in one document, Fascicle VIII.2.

- The OSI specifications are published in two Blue Book documents: Fascicle VIII.4 and VIII.5, in contrast to the Red Book, which used one document.

- The OSI application layer has been "filled in."

- The X.500 Directory Series have been added.

- The X.300 Interworking Series have been expanded.

- The X.400 Message Handling Services have had numerous services added and have changed the structure of some of the common service elements.

Changes Since the Blue Books

The changes made since the 1988 Books have focused on modifying and adding Recommendations. The most significant changes include the "filling-in" of several of a number of "for further study" clauses in the X.300 Recommendations, the addition of documents in most of the categories illustrated in Figure 2.1, and the publication of additional specifications in the X.600 and X.700 Recommendations.

Documentation Tools Used by ITU-T in the X Series

The ITU-T and ISO documents are not written to be tutorials for the consumption of ordinary mortals. They are intended to be technical references for engineers, designers, and programmers. Nonetheless, the specifications include a variety of very helpful documentation tools to depict and explain some rather complex subjects. With one exception, we examine these tools in this chapter; this last tool (ASN.1) is reviewed in chapter 6.

Time sequence diagrams (X.210)

A very useful way to view the communications between a user and a service provider is with a time sequence diagram (see Figure 2.2). The sequence of events takes place in the order of the relative positions of the arrows on the vertical lines; these lines are time lines. In between the time lines is the ser-

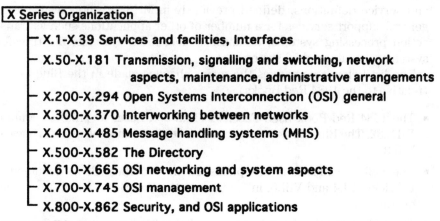

X Series Organization

- X.1-X.39 Services and facilities, Interfaces
- X.50-X.181 Transmission, signalling and switching, network aspects, maintenance, administrative arrangements
- X.200-X.294 Open Systems Interconnection (OSI) general
- X.300-X.370 Interworking between networks
- X.400-X.485 Message handling systems (MHS)
- X.500-X.582 The Directory
- X.610-X.665 OSI networking and system aspects
- X.700-X.745 OSI management
- X.800-X.862 Security, and OSI applications

Figure 2.1 X Series organization.

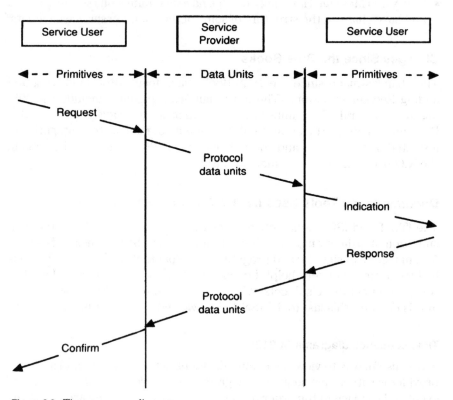

Figure 2.2 Time sequence diagrams.

vice provider, which could be a network, a cell relay switch, a packet switch, etc. The position of the arrows in this figure mean that the confirm primitive is issued as a result of the issuance of the request primitive.

From the context of the X series, the service provider is actually the next lower layer in the OSI protocol stack. Consequently, the service user is the layer above the service provider. The primitives (request, indication, response, confirm) are passed between the upper layer (the service user) and the next lower layer (the service provider). The parameters in the primitives are used to invoke the functions at the next lower layer and to create the appropriate protocol control information (PCI) (more commonly known as headers) to be used by the receiving service user. In this figure, the other service user is shown on the right side of the time sequence diagram.

Figure 2.2 refers to the protocol data units (PDUs). Later chapters examine this term in more detail. For the present the reader should remember it as the self-contained unit of data passed between peer layers, such as a packet, frame, cell, etc. In Figure 2.2 the PDU is created by the rules contained in the protocol specification. The important aspect of this discussion is to remember that the PDU created by one layer in one machine is used by the same (or peer) layer in the other machine to invoke specified operations.

X.210 provides guidance on how to read and interpret time sequence diagrams.

State transition diagrams

Many of the recommendations sponsored by the ITU-T are also explained with state transition diagrams. Figure 2.3 shows the use of such a diagram from the ITU-T X.213 network layer procedures. Each ellipse shows a "state" of a protocol entity instance, such as a process running in a computer. While in a particular state, the protocol entity can issue, receive, and act upon certain primitives. Any other action is logically inconsistent with the intent of the protocol and thus violates the protocol convention. For example, if the entity is in the idle state, it cannot accept data primitives. It can issue and receive connect and disconnect primitives, and then if it successfully sets up a connection and enters the data transfer ready state, it can send and receive data.

State transition diagrams are a graphical tool to show the logic Ω, the protocol, and provide a means to verify the protocol's behavior. They also are quite helpful tools to use when software developers are writing programs for the protocol.

Protocol machines, state tables, event lists, predicates, and action tables

Several of the ITU-T documents describe a protocol with the use of state tables. These tables govern the sequence of events taken by the protocol "ma-

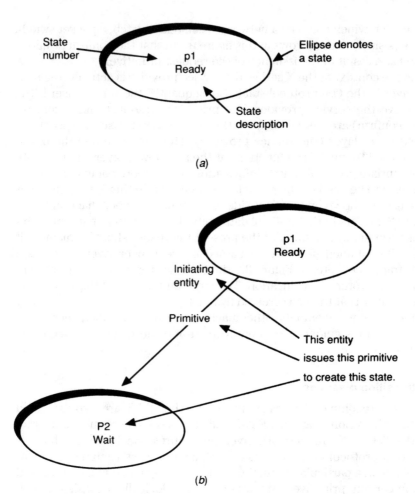

Figure 2.3 State transition diagrams.

chine." The term *machine* is used in the abstract sense to describe a computing entity such as a piece of software, a hardware logic board, etc. To provide an example of a state table, consider the following information taken from a ITU-T OSI applications layer protocol the reliable transfer service element (RTSE), published as X.228:

STA0

RT-OPreq p1:
RTORQ
[a1]
STA01

These cryptic entries contain a wealth of information about certain aspects of the protocol. We will analyze them in the following material.

Assume two protocol machines are to communicate with each other. First, both machines must be in an idle state before they can exchange primitives or PDUs. Further assume that a state table in the ITU-T specification shows that a reliable transfer open primitive is received by the protocol machine from a reliable transfer user. In the above example, this primitive is RT-OPreq, or a Reliable Transfer OPen request primitive.

Because most of the OSI protocols contain many state tables, we first find a table entry with the open primitive as a row entry. As the above example reveals, the primitive is shown to operate on the idle state (STA0) as described in Table 2.2.

TABLE 2.2 Example of a State Table

Table entry	Explanation of entry
STA0	The identification of the state—in this example, state 0. A state description table can be examined to find more information about STA0. It stipulates that the protocol machine is idle and unassociated (not logically connected).
p1:	This entry is called a predicate, and it belongs to yet another table, the predicate table. A predicate is often used to provide supplemental information about the state. In this example, p1 states that the machine in this state can support a requested association (connection) from another machine.
RTORQ	This entry is a reliable transfer open request protocol data unit. It too is used as an entry in another table, an incoming event list. In this example the table entry implies that the source of this PDU is an initiating peer reliable transfer machine (in this example, the machine that received the open primitive from the user) and states that it is received by the acceptor machine as user data within an associate indication primitive. If necessary, a reader can then read about the purpose and contents of this PDU in another section of the OSI manual. This entry in this table means the initiator machine must issue this data unit to the acceptor machine. Another part of the document states that it must be issued as the data parameter in an open primitive to the next lower service provider entity.
[a1:]	This entry is called specific actions. Once again, another table must be read to determine what the actions are. In this example, the a1 action states that the initiating machine is now identified by a boolean condition of True. This condition is used later to determine other actions. Specific action entries also define actions such as the setting of timers, the setting of values in the parameters in primitives, and so forth.
STA01:	This entry describes the resultant state of the association. A lookup to this descriptor in another table (the state descriptor table) reveals that the protocol machine is awaiting a confirm, reject, or abort PDU from the peer machine.

It is fair to say that the ITU-T X Series specifications are not intended to be light reading. They are written to be concise and succinct descriptions of a protocol. However, if someone is really interested in knowing how the protocol works, these documentation tools are invaluable. They are essential reading for the engineer, the designer, and the programmer.

Layer Operations

Several of the ITU-T Recommendations (some of the X Series, the I Series, the Q Series, etc.) are based on layered architectures (introduced in chapter 1). The X Series use several methods for describing the operations between the layers. It is necessary to review some of these operations before delving into the recommendations. While several of these concepts are founded on the OSI Model (the X.200 Series), they are also found in other X Series, such as X.31, and X.25.

Horizontal and vertical communications

OSI protocols and a large number of the X Series specifications allow interaction between functionally paired layers in different locations without affecting other layers. This concept aids in distributing the functions to the layers. In the majority of layered protocols, the data unit, such as a message or packet, that is passed from one layer to another is usually not altered, although the data unit contents may be examined and used to append additional data (trailers and/or headers) to the existing unit.

The relationship of the layers is shown in Figure 2.4. Each layer contains entities that exchange data and provide functions (horizontal communica-

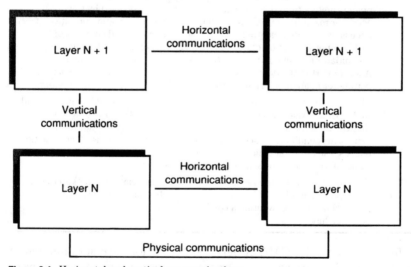

Figure 2.4 Horizontal and vertical communications.

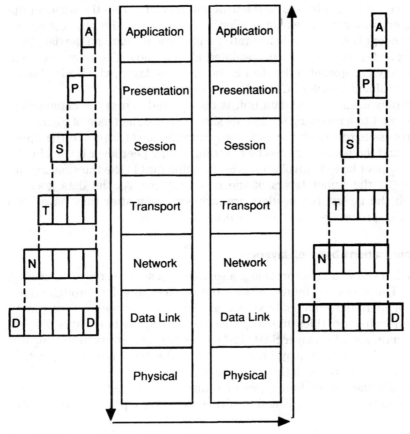

Figure 2.5 Encapsulation and decapsulation.

tions) with peer entities at other computers For example, layer N in one machine communicates logically with layer N in another machine, and the N + 1 layers in the two machines follow the same procedure. Entities in adjacent layers in the same computer interact through the common upper and lower boundaries (vertical communications) by passing parameters to define the interactions.

Encapsulation and decapsulation

Figure 2.5 shows how the layers in and between machines communicate. At a transmitting computer, user data are presented by a user application to the upper layer (application). This layer adds its protocol control information (PCI, which the reader probably knows as a header) to the user data and performs some type of support service to the user. It then passes its header and the user data to the next lower layer, which repeats the process.

Each layer adds a header to the data unit received from the adjacent upper layer (in many systems a header is not added at the physical layer). This concept is somewhat inaccurately called encapsulation (the data from the upper layers are only encapsulated at one end). The only layer that completely encapsulates the data is the data link layer, which adds both a header PCI and a trailer PCI (with some exceptions).

The fully encapsulated data unit is transported across the communications circuit to a receiving station. As shown on the right side of Figure 2.5, the process is reversed; the data goes from the lower layers to the upper layers, and the header created by the transmitting peer layer is used by the receiving peer layer to invoke a service function for (1) the transmitting site and/or (2) the upper layers of the receiving site. As the data goes up through the layers, the headers are stripped away after they have been used. This process is called *decapsulation*.

Communications between layers

Figure 2.6 shows a layer providing a service or a set of services to users A and B. The users communicate with the service provider through an address or identifier commonly known as the *service access point* (SAP). The SAP value identifies an entity that is operating within a layer. For example, a presentation layer SAP (PSAP) identifies an entity in the application layer. The reader may know a SAP by the term *port or socket*. In practice, they accomplish the same thing.

Through the use of four types of transactions, called *primitives* (request, indication, response, and confirm), the service provider coordinates

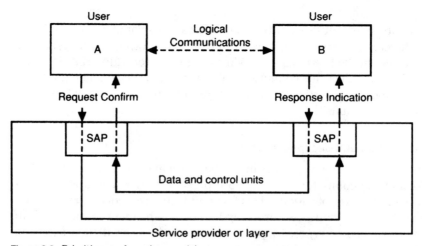

Figure 2.6 Primitives and service provisions.

and manages the communications process between the users (some sessions do not require all primitives):

Request. Primitive by service user to invoke a function

Indication. Primitive by service provider to (1) invoke a function or (2) indicate a function has been invoked at a service access point (SAP)

Response. Primitive by service user to complete a function previously invoked by an indication at that SAP

Confirm. Primitive by service provider to complete a function previously invoked by a request at that SAP

For example, let us assume we wish to establish a *connection* to a network in order to access a computer in another city. Let us also assume we communicate with the network through primitives. We could request the connection with the following primitive: N-CONNECT request (called address, calling address, quality of service parameters, and user data).

This is a request primitive, or more precisely, a network layer (the letter N is for network) connect request primitive. Notice the parameters associated with the connect request. The addresses are used to identify the called and calling parties. The quality of service (QOS) parameters are translated into services for the user; these services are sometimes called *facilities*. They inform the service provider about the type of services that are to be invoked (expedited delivery, for example). It is possible that the primitive could contain many QOS parameters to invoke many facilities. The last parameter, user data, contains the actual data to be sent to the called address. This rather abstract primitive could be coded into a UNIX system call or a C function call.

The primitive is used by the layer to invoke the services within a layer (called *service entities*) and create any headers that will be used by the peer layer entity in the remote station. This point is quite important. The primitives are used by adjacent layers in the local site to create the headers used by peer layers at the remote site.

This brief discourse represents only a small aspect of the use of SAPs. The ITU-T has defined procedures for grouping SAPs, for associating SAPs with names, etc. I have more to say about this topic in later chapters. The interested reader should study X.200 and X.650 if more information is needed about this subject.

Summary

Many of the X Series Recommendations are organized around layered concepts and the OSI Model. The newer X Series Recommendations now make use of OSI service definitions (with primitives) as well as protocol specifications. This chapter has summarized the major documentation tools used by the X Series Recommendations and the relationship to the OSI layers. We also examined the overall structure of the X Series Recommendations and examined the major changes made recently to these standards.

X.1 – X.10

These recommendations are intended to provide a general framework and foundation for the use of common services, facilities, and interfaces by systems that also employ other X Series Recommendations. They also provide some useful definitions. Figure 3.1 depicts the organization of the Services and Facilities Recommendations in relation to the overall X Series Recommendations.

The specific titles of these standards are:

- X.1: International user classes of service in, and categories of access to, public data networks and integrated services digital networks (ISDNs)

- X.2: International data transmission services and optional user facilities in public data networks and ISDNs

- X.3: Packet assembly/disassembly (PAD) facility in a public data network

- X.4: General structure of signals of International Alphabet No. 5 code for character-oriented data transmission over public data networks

- X.5: Facsimile packet assembly/disassembly facility (FPAD) in a public data network

- X.6: Multicast service definition

- X.7: Technical characteristics of data transmission services

- X.10: Categories of access for data terminal equipment (DTE) to public data transmission services

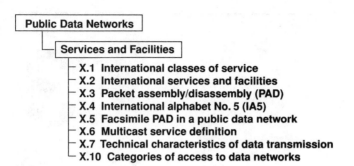

Figure 3.1. The organization of the Services and Facilities Recommendations.

X.1

A user class of service is a category of data communications service in which the call control signalling, the data transfer rates, and the DTE operation modes are identified with a unique number. X.1 specifies these classes of services for various switching and leased technologies. For several years, the ITU-T has published recommendations for the V Series (which define the data signalling rates on the telephone network and modulation rates for modems). X.1 provides a standard for the signalling rates on public data networks and ISDNs.

The X.1 classes of service depend on whether the DTE operates as (1) an asynchronous start-stop device, (2) a synchronous device, or (3) a packet mode device. The user classes of service and the data signalling rates all range from 300 bits per second (bit/s) start-stop asynchronous modes to kilobits per second (64 kbit/s) modes. A vendor or administration may not support all classes of service but typically chooses the appropriate class for a specific need.

X.1 is a very short document, consisting of only a few tables and explanatory notes. These tables often contain references to other ITU-T Recommendations. The tables in X.1 list the user classes of service by number, the data transfer rate, and, if applicable, the call control signals. Table 3.1 provides a partial example of an X.1 table.

TABLE 3.1 Abbreviated Example of an X.1 Class of Service (Service Specific to an ISDN)

User class of service	Data signalling rate
30	64 kbit/s

Note: Call control signals used will be in accordance with those defined for ISDN at S/T.

X.2

Recommendation X.2 is used by a number of other ITU-T Recommendations for the definition of services and facilities. As an example, the X.25 Recommendation makes extensive use of X.2. X.2 standardizes the use of the X.1 classes of service for leased circuits, packet-switched transmission services, and circuit-switched transmission services. As with a number of the ITU-T standards in this chapter, X.2 is principally a list of tables describing the use of facilities on circuit-switched, packet-switched, and leased-circuit systems.

X.2 defines both optional and essential facilities that are made available on a national or international basis. In addition, it stipulates that its facilities are available for an agreed contractual period or on a per-call basis. X.2 also provides guidance for the use of identification methods for DTE data circuit-terminating equipment (DCE) interfaces for X.1 classes.

Table 3.2 shows the facilities available for several services. This table is a summary of several tables in X.2. It does not include explanations on whether the facility is E = Essential, A = Additional, FS = for Further Study, VC = Virtual Call, and PVC = Permanent Virtual Circuit. The reader is encouraged to study X.2 for these more detailed explanations.

TABLE 3.2 X.2 Facilities

Facility	Circuit switch	Packet switch	Customized DTE
Optional user facilities assigned for an agreed contractual period			
Bilateral closed user group	X	X	X
Bilateral closed user group with outgoing access	X	X	X
Calling line identification	X		
Call deflection subscription	X	X	
Call redirection	X	X	
Charging information	X	X	
Closed user group	X	X	X
Closed user group with incoming access	X	X	X
Closed user group with outgoing access	X	X	X
Connect when free	X		
Date and time indication	X		
D-bit modification	X	X	
Default throughout classes assignment	X	X	
Direct call	X	X	X
DTE inactive registration or cancellation	X		
Extended frame sequence numbering	X	X	

TABLE 3.2 X.2 Facilities (Continued)

Facility	Circuit switch	Packet switch	Customized DTE
Optional user facilities assigned for an agreed contractual period			
Extended packet sequence numbering (modulo 128)	X	X	
Fast select acceptance	X	X	
Flow control parameter negotiation	X	X	
Hunt group	X	X	X
Incoming calls barred	X	X	X
Incoming calls barred within a closed user group	X	X	X
Local charging prevention	X	X	
Multilink procedure	X	X	
Nonstandard default packet sizes 16,32,64,256,512,1024,2048,4096	X	X	
Nonstandard default window sizes	X	X	
NUI override	X	X	
NUI subscription	X	X	
Online facility parameter registration or cancellation	X		
Online facility registration	X	X	
One-way logical channel incoming	X	X	
One-way logical channel outgoing	X	X	
Outgoing calls barred	X	X	X
Outgoing calls barred within a closed user group	X	X	X
Packet retransmission	X	X	
Redirection of calls	X		
Reverse charging acceptance	X	X	X
RPOA subscription	X	X	
Throughput class negotiation	X	X	
TOA/NPI address subscription	X	X	
Waiting allowed	X		
Optional user facilities on a per-call basis			
Abbreviated address calling	X	X	X
Bilateral closed user group selection	X	X	
Call deflection selection	X	X	
Call redirection or deflection notification	X	X	
Called line address modified notification	X	X	
Called line identification	X		
Charging information	X	X	X
Closed user group selection	X	X	
Closed user group with outgoing access selection	X	X	
Direct call	X		
Fast select	X	X	
Flow control parameter negotiation	X	X	
Network user identification selection			
NUI selection	X	X	
Multiaddress calling	X		

TABLE 3.2   X.2 Facilities (Continued)

Facility	Circuit switch	Packet switch	Customized DTE
Reverse charging	X	X	X
RPOA selection	X	X	X
Transit delay selection and indication	X	X	
Throughput class negotiation	X	X	

The column labeled "Customized DTEs" warrants further explanation. A packet network may offer customized DTE service to a user in which services are tailored to specific requirements. Typically, the DTE's X.121 address is registered with the network and a DTE "profile" is stored at the network. The profile contains information about the X.2 facilities as well as a number of options for the data link layer (such as timers and retry values).

X.2 also defines the general conventions and procedures for a DTE to identify itself to a network if it is accessing the network through a switched connection, like a telephone dial-up connection. It refers to X.32 for a detailed explanation of the procedure; therefore, we will do the same and defer this discussion until we examine X.32.

Overview of the X.2 facilities for circuit-switched and packet-switched services

This chapter provides an overview of the X.2 facilities. The reader should review the specific ITU-T document (X.25, X.32, etc.) to determine how each is used within each recommendation.

Bilateral closed user groups (BCUGs). This facility allows access restrictions between pairs of DTEs and excludes access to or from other DTEs. The idea is to support bilateral relationships between two DTEs and to allow access between them while, at the same time, excluding access to or from other DTEs.

Bilateral closed user group with outgoing access. This facility allows the DTE to belong to one or more BCUGs and to originate calls into the open part of the network.

Calling line identification. Specifies an international data number based on Recommendation X.121, preceded by **.

Call deflection subscription. The call deflection subscription facility permits the DTE to request an incoming call packet to be deflected to an alternative DTE. The DCE may require the DTE to send this request within a time limit. If the DTE sends the deflection request after the timer expires, the network clears the call.

Call redirection. The call redirection facility redirects calls if the DTE is (1) busy or (2) out of order. Some networks may also provide for systematic call redirection based on other criteria. In addition, some networks may attempt call redirection by (1) accessing a list of alternative DTEs and trying one DTE at a time or (2) chaining the redirections. In the latter case, a call redirected to, say, DTE B could also be redirected to DTE C and then to DTE D and so on.

Charging information. This facility requires the DCE to provide the "charged" DTE information about the session relating to the charges. The facility may be invoked on a per-call basis or as a subscription for a period of time.

Closed user group. The closed user group (CUG) facility provides a level of security or privacy in an "open" network. A DTE can belong to a variable number of CUGs; the limitation is dependent upon the network.

Closed user group with incoming access. A DTE will receive calls from DTEs belonging to the open part of the network (open means a non-CUG) and from DTEs, which are members of other CUGs with outgoing access.

Closed user group with outgoing access. A DTE may initiate calls to all DTEs belonging to the open part of the network and to DTEs, which are members of other CUGs with incoming access. If the DTE has a preferential CUG, only the closed user group selection facility can be used at the DTE-DCE interface.

Connect when free. Allows a device to inform the network that it can obtain and hold a connection when resources are noted as available by the network.

Date and time indication. This facility is used to inform the network user the date and time that the call is established.

D-bit modification. The D-bit modification facility is intended for use by DTEs operating on networks that support end-to-end acknowledgment procedures developed prior to the introduction of the 1980 D-bit procedure. It allows these DTEs to continue to obtain this service *within* the network. This means an acknowledgment value in a packet has end-to-end significance within the network.

Default throughput classes assignment. This facility provides for the selection of one of the throughput rates (in bit/s). *Throughput* describes the maximum amount of data that can be sent through the network when the network is operating at saturation. Factors that influence throughput are line speeds, window sizes, and the number of active sessions in the network.

Direct call. Allows a DTE to make a call to the network without going through the preliminary overhead of a call establishment. For example, in X.21, the DTE can bypass the selection signals phase (state 4) to obtain a connection.

DTE inactive registration or cancellation. The facility allows the DTE to inform the network that a period of time exists in which the DTE is not able to accept incoming calls.

Extended frame sequence numbering. This facility allows the data link layer to use extended sequencing. In effect, the sequencing can range from 0 to 127. The facility is useful for communications links that experience long propagation delays and/or fast throughput.

Extended packet sequence numbering (modulo 128). This facility provides packet sequence numbering using modulo 128 (sequence numbers 0 to 127) for all channels at the DTE-DCE interface. In its absence, sequencing is done with modulo 8 (sequence numbers 0 to 7).

Fast select acceptance. This facility allows a call request to contain user data of up to 128 octets. The called DTE is allowed to respond with a call accepted packet, which can also contain user data. The call request or incoming call indicates whether the remote DTE is to respond with clear request or call accepted signal.

Flow control parameter negotiation. This facility allows the window and packet sizes of a packet session to be negotiated on a per-call basis for each direction of transmission. In many networks, the DTE negotiates packet sizes and window sizes during the call setup. The called DTE (if it subscribes to these facilities) may reply with a counterproposal. If it does not, it is assumed that the call setup parameters are acceptable.

Hunt group. This facility distributes incoming calls across a designated grouping of DTE-DCE interfaces. This 1984 addition gives users the ability to allocate multiple ports on a front-end processor or computer. These multiple ports are managed by the DCE, which is responsible for distributing the calls across them. The manner in which they are distributed is not defined in the X.2 Recommendation.

Incoming calls barred. This facility prevents incoming calls from being presented to the DTE. It applies to all logical channels at the DTE-DCE interface and cannot be changed on a per- call basis. A DTE subscribing to incoming calls barred can initiate calls but cannot accept them.

Incoming calls barred within a closed user group. A DTE may initiate calls to other members of the CUG but cannot receive calls from them. This facility is equivalent to establishing all logical channels as one-way outgoing (originate only).

Local charging prevention. This facility authorizes the DCE to prevent the establishment of calls for which the subscribing DTE would ordinarily pay. The DCE can prevent the calls from taking place by not passing them to the

DTE. The DCE may also charge yet another party (e.g., identify the third party by the NUI facility).

Multilink procedure. This facility permits the use of more than one data link between computers and/or switches. It is useful for increasing through-put between two machines. It is also used to obtain backup to a single link.

Nonstandard default packet sizes. This facility provides for the selection of nonstandard default packet sizes. The default size is 128 octets of user data in most packet networks. The size can be different for each direction of data flow. Some networks that allow different levels of priorities will also allow different packet sizes for different priority calls.

Nonstandard default window sizes. This facility allows the packet transmit window sizes of a DTE to be expanded beyond a default value. It is possible for the window sizes to be different at each end of the connection, and some networks constrain the default window size to be the same for each direction of transmission across the DTE-DCE interface.

NUI override. This facility is established to override the subscription-time facilities. The override only pertains to the specific call. A network usually places restrictions on which subscription-time facilities may be associated with the NUI and the NUI override facility. The reader should use this feature *vis-a-vis* a specific network.

NUI subscription. This facility allows the DTE to furnish the network information on billing, security, or management matters. The information is contained in either the call request or the call accepted packet. It may be used regardless of any subscription to the local charging prevention facility.

Online facility parameter registration or cancellation and online facility registration. These facilities permit the local DTE to request facilities or to obtain the parameters (values) of the facilities. The local DCE returns the current value of all the facilities applicable to the DTE-DCE interface. The facility can be canceled also.

One-way logical channel outgoing and one-way logical channel incoming. These two facilities restrict a logical channel to originating calls only or receiving calls only. They are set when the user subscribes to the network and cannot be changed on a per-call basis. These facilities provide more specific control than the calls barred facilities because they operate on a specific channel or channels.

Outgoing calls barred. This facility prevents the DCE from accepting outgoing calls from the DTE. It applies to all logical channels at the DTE-DCE interface and cannot be changed on a per-call basis. DTE subscribing to outgoing calls barred can receive calls but cannot initiate them.

Outgoing calls barred within a closed user group. A DTE may receive calls from other members of the CUG but cannot initiate calls to them. This facility is equivalent to establishing all logical channels as one-way incoming (terminate only).

Packet retransmission. This facility applies to all logical channels at the DTE-DCE interface. A DTE (but not the DCE) may request retransmission of one to several data packets. The DTE specifies the logical channel number and a value for P(R) in a reject packet. The DCE must then retransmit all packets from P(R) to the next packet that it is to transmit for the first time.

Redirection of calls. The originally called DTE (remote) does not receive an incoming call packet when the redirection is performed. It is the responsibility of the remote DCE to redirect the call. See call deflection for another variation of this service.

Reverse charging acceptance. The reverse charging acceptance facility authorizes the remote DCE to pass to the DTE the incoming calls which request the reverse charging. Otherwise, the DCE will not pass the calls, and the originating DTE will receive a clear indication signal.

RPOA subscription and selection. These facilities allow a calling DTE to specify one or more recognized private operating agencies (RPOA) to handle the packet session. The RPOA is a packet network carrier (a value-added carrier), and it acts as a transit network within one country or between countries. The *RPOA subscription* facility is used with all virtual calls involving more than one RPOA and one or more gateways. The *RPOA selection* facility is used for an individual virtual call, and it is not necessary to subscribe to the RPOA selection facility to use this facility. With these facilities, the call request contains a sequence of RPOA transit networks to handle the call. The call is set up and the traffic is routed based on this information.

Throughput class negotiation. This facility allows the throughput rates to be negotiated on a *per-call* basis. A throughput greater than a DTE's default values is not allowed. The allowable rates that can be negotiated are 75, 150, 300, 600, 1200, 2400, 4800, 9600, 19,200, 48,000, and 64,000 bit/s, although many vendors support higher speeds.

TOA/NPI address subscription. This facility supports extended address fields in the call setup and clearing packets. It was added in the 1988 Blue Book to permit the use of an extended E.164 ISDN addressing scheme. When this facility is in use, the DCE uses the long address format, but the DTE has the option of either using or not using it.

Waiting allowed. In circuit-switched networks, this facility permits the DTE to be given wait periods to complete a call with another DTE.

X.3

The idea of a PAD is to provide protocol conversion for a user device (DTE) to a public or private packet data network and a complementary protocol conversion at the receiving end of the network. The goal is to provide a transparent service to user DTEs through the network. The asynchronous-oriented PAD performs the following functions:

- Assembly of characters into packets
- Disassembly of user data field at other end
- Handling virtual call setup, clearing, resetting, and interrupt packets
- Generation of service signals
- Mechanism for forwarding packets (a full packet or when a timer expires)
- Editing PAD commands
- Automatic detection of data rate, code, parity, and operational characteristics

The 1988 version of X.3 defines a set of 22 parameters the PAD uses to identify and control each terminal communicating with it. When a DTE-DCE connection is established, the PAD parameters are used to determine how the PAD communicates with the user DTE and vice versa. The user DTE also has the option of altering the parameters after the logon to the PAD device is complete. Figure 3.2 shows a typical PAD configuration.

Each of the 22 parameters consists of a reference number and parameter values. These parameters and references are explained in general terms in Table 3.3. The reader should refer to the specific ITU-T Recommendation for more detailed information on the use of these parameters.

X.4

This brief recommendation describes some conventions for the use of the IA5 code on data transmission networks. Essentially, it defines the conventions for the order of the transmission of the bits and the stipulation for the use of parity bits for IA5. It assumes the transmission is with character-oriented protocols.

X.5

X.5 was published in 1991 to establish procedures for the support of facsimile transmission through a public data network. As the reader may know, the ITU-T term of public data network is meant to delineate it from a general switched telephone network (GSTN). Most readers of this book are likely more familiar with transmitting facsimile over a GSTN, more commonly

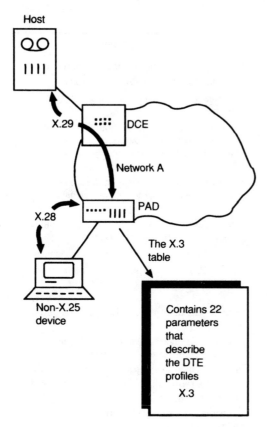

Figure 3.2 The X.3, X.28, and X.29 Recommendations.

known as POTS (plain old telephone system). Therefore, this recommendation provides guidance on transmitting a fax through the conventional dial-up lines on the telephone network to a local packet assembler/disassembler, then through the packet network, next to a remote packet assembler/disassembler, and finally to the telephone network to the end user device.

The typology for X.5 is quite similar to that of X.32 which was illustrated in Figure 3.2. The PAD in Recommendation X.5 is identified as the facility pad (FPAD).

As with many of the ITU-T Recommendations, X.5 is a very general document which sets the overall framework for the function of the FPAD with the user ITU-T G3 facsimile equipment. It defers to Recommendations X.38 and X.39 for a more detailed explanation of the operations of the FPAD. The principal use of X.5 is the definition of one FPAD parameter which is exchanged between the user DTE and the FPAD. This parameter is simply identified as the control of FPAD service signal. Its function is to provide the G3 fax equipment with information on the format of FPAD service signals.

TABLE 3.3   The X.3 PAD Parameters

X.3 parameter reference number	Description
1—Pad recall	Escape from data transfer mode to command mode in order to send PAD commands
2—Echo	Controls the echo of characters sent by the terminal
3—Data forwarding	Defines the characters to be interpreted by the PAD as a signal to forward data; indication to complete assembly and forward a complete packet
4—Idle timer delay	Selects a time interval between successive characters of terminal activity as a signal to forward data
5—Ancillary device control	Allows the PAD to control the flow of terminal data using XON/XOFF characters
6—Control of PAD service signals	Allows the terminal to receive PAD messages
7—Operation of the PAD on receipt of breaking signal from DTE	Defines PAD action when a break signal is received from the terminal
8—Discard output	Controls the discarding of data pending output to a terminal
9—Padding after carriage	Controls PAD insertion of padding characters after a carriage return is sent to the terminal
10—Line folding	Specifies whether the PAD should fold the output line to the terminal; predetermined number of characters per line
11—Binary speed of DTE	Indicates the speed of the terminal; cannot be changed by DTE
12—Flow control of the PAD	Allows the terminal to flow control data being transmitted by the PAD
13—Line feed insertion	Controls PAD insertion of line feed after a carriage return is sent to the terminal
14—Line feed padding	Controls PAD insertion of padding characters after a line feed is sent to the terminal
15—Editing	Controls whether editing by PAD is available during data transfer mode (parameters 16, 17, and 18)
16—Character delete	Selects character used to signal character delete
17—Line delete	Selects character used to signal line delete
18—Line display	Selects character used to signal line display
19—Editing PAD service signals	Controls the format of the editing PAD service signals
20—Echo mask	Selects the characters which are not echoed to the terminal when echo (parameter 2) is enabled
21—Parity treatment	Controls the checking and generation of parity on characters from and to the terminal
22—Page wait	Specifies the number of lines to be displayed at one time

Because X.5 is a general recommendation that is intended to establish the framework for the use of facsimile over data networks, we shall turn our attention later to its companion specifications, X.38 and X.39.

X.6 and X.7

X.6 and X.7 are new recommendations dealing with various aspects of multicast service definition, and additional technical characteristics were not available as of the writing of this book.

X.10

This recommendation was approved for the 1984 ITU-T Red Books. X.10 defines the different categories of access for DTEs into different kinds of networks. Specifically, the standard defines how DTEs interface into (1) circuit-switched networks, (2) packet-switched networks, and (3) leased-circuit networks. In addition, the standard defines how terminals interface into an ISDN. X.10 also stipulates the data signalling rate and the physical interface requirements (such as X.21, X.21 bis, etc.). The document contains many tables. We will include one of them here as Table 3.4 to give the reader an idea of the structure of X.10.

TABLE 3.4 Synchronous Switched Access by Means of an ISDN B Channel to a Packet-Switched Data Transmission Service (Abbreviated Sample)

Category of access	Signalling rate	DTE-DCE interface requirements
Q1	2,400bit/s	See Recommendations
Q2	4,800bit/s	X.31 and X.32,
Q3	9,600bit/s	depending on use of
Q4	48,000bit/s	S/T or R
Q5	64,000bit/s	

Summary

The X Series services and facilities recommendations are fairly terse descriptions of several operations in circuit-switched and packet-switched networks. The widely used X.3 PAD specification is included in this group of recommendations, as is the very useful X.2 Recommendation that deals with facility descriptions. This chapter provided a brief summary of these recommendations. Many of these recommendations are used in other X Series Recommendations discussed in subsequent chapters.

4

X.20 – X.39

The ITU-T X Series Data Communication Network Interfaces Recommendations cover several of the layers of a typical OSI protocol suite. The majority of the specifications define operations at either the physical or network layer. The general organization of these recommendations is shown in Figure 4.1. The specifications' titles are:

- X.20: Interface between data terminal equipment (DTE) and data circuit-terminating equipment (DCE) for start-stop transmission services on public data networks

- X.20 bis: Use on public data networks of data terminal equipment (DTE) which is designed for interfacing to asynchronous duplex V-Series modems

- X.21: Interface between data terminal equipment (DTE) and data circuit-terminating equipment (DCE) for synchronous operation on public data networks

- X.21 bis: Use on public data networks of data terminal equipment (DTE) which is designed for interfacing to synchronous duplex V-Series modems

- X.22: Multiplex DTE/DCE interface for user classes 36

- X.24: List of definitions for interchange circuits between data terminal equipment (DTE) and data circuit-terminating equipment (DCE) on public data networks

- X.25: Interface between data terminal equipment (DTE) and data circuit-terminating equipment (DCE) for terminals operating in the packet mode and connected to public data networks by dedicated circuit

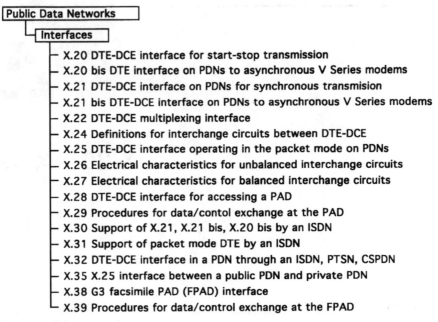

| Public Data Networks |
| Interfaces |

- X.20 DTE-DCE interface for start-stop transmission
- X.20 bis DTE interface on PDNs to asynchronous V Series modems
- X.21 DTE-DCE interface on PDNs for synchronous transmision
- X.21 bis DTE-DCE interface on PDNs to asynchronous V Series modems
- X.22 DTE-DCE multiplexing interface
- X.24 Definitions for interchange circuits between DTE-DCE
- X.25 DTE-DCE interface operating in the packet mode on PDNs
- X.26 Electrical characteristics for unbalanced interchange circuits
- X.27 Electrical characteristics for balanced interchange circuits
- X.28 DTE-DCE interface for accessing a PAD
- X.29 Procedures for data/contol exchange at the PAD
- X.30 Support of X.21, X.21 bis, X.20 bis by an ISDN
- X.31 Support of packet mode DTE by an ISDN
- X.32 DTE-DCE interface in a PDN through an ISDN, PTSN, CSPDN
- X.35 X.25 interface between a public PDN and private PDN
- X.38 G3 facsimile PAD (FPAD) interface
- X.39 Procedures for data/control exchange at the FPAD

Figure 4.1 X Series interfaces.

- X.26: Electrical characteristics for unbalanced double-current interchange circuits for general use with integrated circuit equipment in the field of data communications

- X.27: Electrical characteristics for balanced double-current interchange circuits for general use with integrated circuit equipment in the field of data communications

- X.28: DTE/DCE interface for a start/stop mode data terminal equipment accessing the packet assembly/disassembly facility (PAD) in a public data network situated in the same country

- X.29: Procedures for the exchange of control information and user data between a packet assembly/disassembly (PAD) facility and a packet mode DTE or another PAD

- X.30: Support of X.21 and X.21 bis based data terminal equipments (DTEs) by an integrated services digital network (ISDN)

- X.31: Support of packet mode terminal equipment by an ISDN

- X.32: Interface between data terminal equipment (DTE) and data circuit-terminating equipment (DCE) for terminals operating in the packet mode and accessing a packet switched public data network through a public switched telephone network or an integrated services digital network or a circuit switched public data network

- X.35: Interface between a PSPDN and a private PSDN which is based on X.25 procedures and enhancements to define a gateway function that is provided in the PSPDN
- X.38: G3 facsimile equipment/DCE interface for G3 facsimile equipment accessing the facsimile packet assembly/disassembly facility (FPAD) in a public data network situated in the same country
- X.39: Procedures for the exchange of control information and user data between a facsimile packet assembly/disassembly (FPAD) facility and a packet mode data terminal equipment (DTE) or another FPAD

Physical Layer Interfaces

We learned earlier that the ITU-T publishes the V Series physical layer protocols to include (1) the DTE-to-DCE interface and (2) the DCE-to-DCE interfaces for the telephone network. Several of the X Series Recommendations provide the same type of definitions and conventions for data networks (as usual, ITU-T documents invariably use the term "public data networks," but these standards are applicable to private networks as well).

X.24

X.24 is analogous to V.24 for the V Series in that it provides the descriptions of the interchange circuits used by the X Series interfaces, just as V.24 provides the descriptions of the interchange circuits used by the V Series interfaces. The ITU-T uses X.24 to define the transfer of data, call control signals, and timing signals between the DTE and the DCE. Tables 4.1 and 4.2 show how the X.24 interchange circuits are used and designate them as data, control, or timing circuits.

TABLE 4.1 X.24 Interchange Circuits

		Data		Control		Timing	
	Interchange circuit	From	To	From	To	From	To
G	Signal ground or common return						
Ga	DTE common return				X		
Gb	DCE common return			X			
T	Transmit		X		X		
R	Receive	X		X			
C	Control				X		
I	Indication			X			
S	Signal element timing					X	
B	Byte timing					X	
F	Frame start identification					X	
X	DTE signal element timing						X

TABLE 4.2 X.24 Interchange Circuits Legend

Circuit G Signal ground or common return	Establishes the signal common reference potential for the interchange circuits.
Circuit Ga DTE common return	Connected to the DTE circuit common and used as the reference for the unbalanced X.26-type interchange circuit receivers within the DCE.
Circuit Gb DCE common return	Connected to the DCE circuit common and used as the reference for the unbalanced X.26-type interchange circuit receivers within the DTE.
Circuit T Transmit	Signals originated by the DTE are transferred on this circuit to the DCE.
Circuit R Receive	Signals sent by remote DTE are transferred on this circuit to the DTE.
Circuit C Control	This circuit controls the DCE for a particular signalling process, depending upon the use of the relevant recommendation for the procedural operations at the interface. Representation of a call control signal requires additional coding of the T circuit depending on the use of the specific procedural recommendation. During data transfer, this circuit is on.
Circuit I Indication	Indicates to the DTE the state of the call control process. Representation of a call control signal requires additional coding of circuit R, depending upon the procedural recommendation.
Circuit S Signal element timing	Provides the DTE with signal element timing information.
Circuit B Byte timing	Provides the DTE with 8-bit byte timing information.
Circuit F Frame start identification	Continuously provides the DTE with a multiplex frame start indication when connected to a multiplexed DTE-DCE interface.
Circuit X DTE transmit signal element timing	Provides signal element timing information for the transmit direction in cases where circuit S only provides signal element timing for the receive direction.

One might question why two interchange descriptions are needed. After all, a transmit data interchange circuit transmits the same data out of a DTE using a V.24 interface as it does with an interface using an X.24 interface.

Fewer interchange circuits are defined in X.24, because the X.24 receive circuit (R) and transmit circuit (T) are also used to transfer call control signals originated by the DTE and DCE. In contrast, the majority of the V.24 circuits are separate control circuits. It probably is not stretching the point to categorize the X.24 Recommendation as an in-band interface signalling interface standard in contrast to the V.24 Recommendation, which could be classified as an out-of-band signalling interface standard.

The X.24 Recommendation uses interchange circuits T, R, C, and I for the transfer of data and control signals. The four circuits are activated with a variety of signals to convey different types of operations. For example, the T and R interchange circuits convey binary 1s and 0s; and the C and I circuits are set to on or off, in various combinations, to convey signals such as DTE ready, DCE ready, incoming call, etc. The 1, 0, off, and on signals are not the only signals that can be conveyed. For example, the R and T circuits can convey ASCII-type characters such as plus signs, SYN signals, and other kinds of characters to provide additional signalling information.

As we just learned, the V Series counterpart to X.24 is V.24. V.24 is considered inadequate to handle the more complex call-control functions found in a data network and ITU-T states that V.24 is inappropriate for signalling across a data network interface. Therefore, a number of X.24 functions deal with (1) call-control functions, such as call request, incoming call, and call clear, and (2) selection of facilities. The selection of facilities is in accordance with Recommendation X.2, discussed in chapter 3.

Other X Series Physical Layer Interfaces

Several X Series physical layer interfaces are employed for the following functions:

- X.20: Interface between DTE and DCE for start/stop transmission services on public data networks.

- X.22: Multiplex DTE/DCE interface for user classes 36.

- X.20 bis: Used on public data networks with data terminal equipment (DTE) which is designed for interfacing to asynchronous duplex V Series modems.

- X.26: Electrical characteristics for unbalanced double-current interchange circuits for general use with integrated circuit equipment. The text of V.10 is used for this recommendation.

- X.27: Electrical characteristics for balanced double-current interchange circuits for general use with integrated circuit equipment. The text of V.11 is used for this recommendation.

X.26 and V.10

As stated above, these recommendations define interchange circuits for unbalanced systems. Unbalanced in this context means that each interface circuit has one conductor. These conductors share a common ground, and the voltage signals are measured in relation to the polarity of the voltage with respect to ground. The interfaces are intended for conventional wire interfaces between the DTE and the DCE, but with modifications, they can be used with coaxial cable. X.26 defers to the text in V.10 to describe this interface.

V.10 is similar to V.28 (which is often used with X.21 bis). The principal difference relates to the bit rate (bit/s) across the interface. V.10 permits a data rate of up to 100 kbit/s. Also, the voltage thresholds for V.10 are +0.3 and 0.3 volts (V). ITU-T recommends that a V.10 cable not exceed 10 meters (m). However, users have extended this distance with satisfactory results.

X.27 and V.11

X.27 and V.11 specify a balanced connection. In this regard, a balanced circuit has two conductors, a signal and a return. The measurement of the voltage pertains to the difference between the signals on these two conductors. X.27 defers to the text in V.11 to describe this interface.

In addition to the V.11 interface being balanced, the two wires associated with each interchange circuit carry signals as follows. First, one wire carries the original true form of the signal while the other wire carries an inverted coding of the signal. This technique is known as differential signalling and works quite well on noisy lines because the original signal can be recovered by taking the mathematical difference between the received original signal and the received inverted signal.

V.11 interfaces are usually implemented with twisted pair cable. The wires carry the A and B forms of the signal discussed in the previous paragraph. This is a useful technique because the electrical noise that is coupled into the wire is coupled equally into both wires on the interchange circuit. As just mentioned, the logic at the receiver allows the noise to be canceled out on each wire.

As with V.10, the V.11 receiver thresholds are +0.3 and 0.3 V. Remember that the difference in V.11 is that the received voltage has a difference between signal A and B, whereas with V.10 it represents the difference relative to ground.

X.21

The X.21 Recommendation is yet another interface standard that has received considerable attention in the industry. It is used in several European countries and Japan but has seen limited implementation in North America.

X.21 is designed to support different signalling rates. For rates of 9600 bit/s and below, the DCE side of the interface complies with X.27 and the DTE side complies with X.27 or X.26. For rates of 9600 bit/s and above, both the DTE and the DCE sides comply with X.27. The X.21 interface uses the ISO 4903 14-pin connector.

X.21 employs the X.24 interchange circuits in accordance with the rules shown in Table 4.3. The T and R circuits transmit and receive data across the interface. Unlike EIA-232-E and the V Series, X.21 uses the T and R circuits for transferring user data as well as control signals.

TABLE 4.3 X.21 Interchange Circuits

Interchange circuit designation	Interchange circuit name	Direction	
		To DCE	From DCE
G	Signal ground or common return		
Ga	DTE common return	X	
T	Transmit	X	
R	Receive		X
C	Control	X	
I	Indication		X
S	Signal element timing		X
X	DTE signal element timing	X	
B	Byte timing		X

The C circuit provides control signals to the DCE and the I circuit provides the indication signals to the DTE. These two circuits are used to activate, manage, and deactivate the DTE-DCE interface session. The S and B circuits provide for signals to synchronize the signals between the DTE and DCE. The G circuit acts as a signal ground or a common return.

The X.21 states

X.21 is designed around the concepts of states and state diagrams discussed in chapter 2. The X.21 states are shown in Table 4.4 (the state diagrams are provided in Annex A of the X.21 Recommendation). The recommendation places strict rules on how the X.24 interchange circuits are operated to enter and leave the states. Various combinations of signals are used to create these states and acquire other services from the network, such as facilities (see X.2), status information, and error indications. Moreover, X.21 defines the use of over 20 timer values to govern the state transitions and the actions that occur in the event the time limits on the timers expire. The next section provides a detailed example of the use of these states to create a connection with a network.

Example of X.21 operations

Figures 4.2 and 4.3 provide an example of the operations between the DTE and the DCE in obtaining a connection with the network, transferring data, and releasing the connection. Figure 4.2 shows the operations involved in creating and clearing a connection at the local end and Figure 4.3 shows the operations at the remote end. Each of the events (numbered 1 through 16 in the figures) is described. A number of important X.21 features associated with the events are examined. Figure 4.2 includes the event numbers that relate to the operations in Figure 4.3. The reader should be able to easily "fill-in" the events in Figure 4.3 by referring to Figure 4.2. Several (but not all) of the X.21 timers are also brought into this discussion. The signals off and on are represented on the X.21 interface with continuous 1s and 0s respectively.

The reader will find Table 4.4 helpful during this discussion.

TABLE 4.4   X.21 States

State number	State name
1	Ready
2	Call request
3	Proceed to select
4	Selection signal sequence
5	DTE waiting
6A	DCE waiting
6B	DCE waiting
7	DCE provided information (call progress signals)
8	Incoming call
9	Call accepted
10A	DCE provided information (to calling DTE)
10B	DCE provided information (to calling DTE)
10C	Call information signal sequence state
11	Connection in progress
12	Ready for data
13	Data transfer
13S	Send data
13R	Receive data
14	DTE controlled not ready, DCE ready
15	Call collision
16	DTE clear request
17	DCE clear confirmation
18	DTE ready, DCE not ready
19	DCE clear indication
20	DTE clear confirmation
21	DCE ready
22	DTE uncontrolled not ready, DCE not ready
23	DTE controlled not ready, DCE not ready
24	DTE uncontrolled not ready, DCE ready
25	DTE-provided information

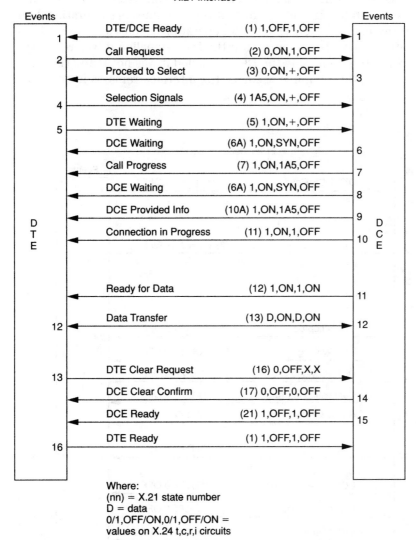

Figure 4.2 X.21 local operations.

Event 1. The DTE and DCE are in state 1 (DTE-DCE Ready). This state is identified with the DTE signalling t = 1 and c = off and the DCE signalling r = 1, and i = off. These signals indicate that the user and the network are ready to establish a connection by entering into operational phases. The ready state may pass to either the 14, 22, 23, or 24 states. These states mean that either the DTE or DCE is unable to enter into a connection setup because of problems (states 22 and 23) or that the machines are operational but temporarily unable to enter into the operational phases (states 14 and 24). State 18 may also be entered if the DTE signalled it was ready (t = 1, c = off) at the same time the DCE signalled it was not ready (r = 0, i = off). However, this example assumes the interface remains in the ready state.

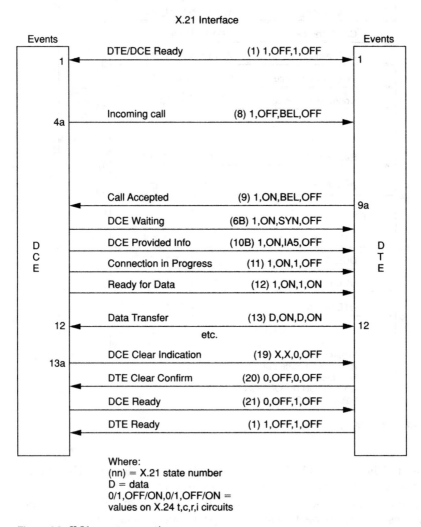

Figure 4.3 X.21 remote operations.

Event 2. The DTE sends a call request signal to the network with t = 0 and c = on. Upon sending this signal, the T1 DTE timer is started (which is called a DTE time limit). If T1 expires before receiving back an appropriate signal from the DCE, the DTE must enter into DTE ready (state 1).

Event 3. Assuming the network is willing to accept a selection signal from the DTE, the DCE enters state 3 with r = +(continuous IA5 + characters) and i = off. Although not shown in Figure 4.2, the DCE sends two or more contiguous SYN characters before the +signals. Upon sending these signals to the DTE, the DCE starts timer T11 (which is called a DCE time-out). If T11 expires before receiving the appropriate signal from the DTE (an end-of-selection signal, discussed shortly), several remedial operations may take place (discussed in Annex C of X.21). The DTE's T1 timer is terminated by receiving the proceed to select signal from the DCE.

Event 4. The selection signal phase begins with the DTE placing c = on and sending a specific order of signals across the t interchange circuit, the first of which are two or more contiguous SYN characters, which are depicted in the figure with the notation "IA5," for International Alphabet Number 5 characters. The selection signals consist of several orderings of the following "blocks":

Facility request block: Contains information regarding the X.2 facilities, such as call redirection, charging information, etc. The reader should be aware that this part of X.21 does not correlate well to X.2. The reader is encouraged to check this feature carefully in the vendor product.

Address block: Contains the X.121 address(es) of the called party (parties).

Facility registration block: In the event the DTE wishes to register facilities, they are coded in this block.

Facility cancellation block: Any cancellation of facilities is coded in this block.

End of selection signal: Informs the DCE that all selection signals have been transmitted, and the DCE terminates its T11 timer. This signal consists of t = +character.

Event 4A. The reception of these signals is sufficient for the network to establish the call, the result of which is shown in Figure 4.3 as event 4A, the incoming call signal from the remote DCE to the remote DTE. The IA5 BEL character is sent across r and i = off.

Event 5. After sending the selection signals, the DTE enters the DTE waiting state by signalling t = 1, c = on. At this time it starts the DTE T2 timer.

Event 6. The DCE receives the DTE signals and enters the DCE waiting state by sending two or more contiguous SYNs on interchange circuit r and by setting i = off.

Event 7. The call progress phase is entered by the DCE signalling a call progress signal of IA5 characters on interchange circuit r and setting i = off. The call progress signal informs the DTE about the progress (or lack of progress) of the connection setup. X.21 defines the coding for this signal, but the definition is provided in X.96 (see chapter 5). X.21 allows the T2 timer to be terminated upon receipt of this signal. This timer may also be terminated by receiving the following signals:

DCE-provided information: State 10A
Ready for data: State 12
DCE clear indication: State 19

Event 8. After the call progress signal is transmitted to the DCE, the network enters into the DCE waiting state by signalling two or more SYNs on interchange circuit r and setting i = off.

Event 9. The DCE-provided information state (10A) is used to provide the DTE with information about several aspects of the call, such as charging information, time of call establishment, requested facilities, closed user groups, etc.

Event 9A. The remote DTE accepts the call in Figure 4.3 by sending its DCE a call accepted signal, with t = 1 and c = on.

Event 10. The connection in progress informs the DTE that the call is underway by signalling r = 1 and i = off.

Event 11. After the receiving (remote) DTE signals it is ready for data transfer (by sending a call accepted signal to its remote DCE, see Figure 4.2), the local DCE enters the ready for data state with r = 1 and i = on.

Event 12. The DTE and DCE can now exchange data. The notation of "D" in the figures means DTE and DCE data signals. Note that c = on and i = on during the data transfer. The DTE signals an end of data transfer with t = 1, c = off; the DCE signals an end of data transfer with r = 1, i = off, which are not shown in the figures.

Event 13. The local DTE chooses to terminate the connection by signalling a clear request with t = 0 and c = off. Upon sending this signal, it starts timer T5. The network or the remote DTE also has the option of terminating this call, but these options are not depicted in this example. The X values on r and i mean that X.21 permits several variations of signals on these circuits for state 16.

Event 13A. The effect of the clear request in Figure 4.3 is a clear indication at the remote interface. The DCE provides this signal with r = 0 and i = off.

Event 14. The DCE responds with the DCE clear confirm signal with r = 0 and i = off. The DTE T5 timer many be terminated by receiving this signal or upon receiving the DCE ready state (event 15).

Event 15. The DCE places its interface into state 21 (DCE ready) by signalling DCE ready with r = 1, i = off.

Event 16. The DTE places its interface into state 1 (DTE ready) by signalling DTE ready with t = 1, c = off.

X.21 bis

This recommendation was published by the ITU-T to serve as an interim standard for using synchronous transmission schemes to interface the V Series modems with public packet networks. X.21 was intended to eventually replace X.21 bis, but this has not happened. X.21 bis is often used as the physical interface in an X.25 packet network. The Blue Book recommends X.21 bis be used to support the following call management services (see Table 4.5):

- Leased point-to-point services and centralized multipoint services
- The direct call facility
- The address call facility

X.21 bis uses the V.24 circuits. It also has several ways to use the ISO connectors and the other V and X interfaces. The electrical characteristics of the interchange circuits at both the DCE and the DTE sides of the interface may comply either with Recommendation V.28 using the 24-pin connector and ISO 2110 or with Recommendation X.26 using the 37-pin connector and ISO 4902. In North America, the V.28 convention is often used.

TABLE 4.5 X.21 bis Call Management

V.24 interchange circuit	Condition	Meaning
107	on	Ready for data
107	off	DCE clear indication
107	off	DCE clear confirmation
108/1	on	Call request
108/1	on	Call accepted
108/1	off	DTE clear request
108/1	off	DTE clear confirmation
108/2	on	Call accepted
108/2	off	DTE clear request
108/2	off	DTE clear confirmation

X.25

The ITU-T issued the first draft of X.25 in 1974. It was revised in 1976, 1978, 1980, 1984, and last in 1988. Since 1974, the standard has been expanded to include many options, services, and facilities, and several of the newer OSI protocols and service definitions operate with X.25. X.25 is now the predominant interface standard for wide area packet networks.

The placement of X.25 in packet networks is widely misunderstood. X.25 is *not* a packet-switching specification. It is a packet network interface specification (see Figure 4.4). X.25 says nothing about the routing within the network. Hence, from the perspective of X.25, the network is a "cloud." For example, the X.25 logic is not aware if the network uses adaptive or fixed directory routing. The reader may have heard of the term *network cloud.* Its origin is derived from these concepts. X.25 defines the procedures for the exchange of data between a user device (DTE) and the network (DCE). Its formal title is "Interface between Data Terminal Equipment and Data Circuit Terminating Equipment for Terminals Operating in the Packet Node on Public Data Networks." In X.25, the DCE is the "agent" of the packet network to the DTE.

X.25 establishes the procedures for two packet-mode DTEs to communicate with each other through a network. It defines the two DTEs sessions with their respective DCEs. The idea of X.25 is to provide common procedures between a user station and a packet network (DCE) for establishing a session and exchanging data. The procedures include functions such as identifying the packets of specific user terminals or computers, acknowledging packets, rejecting packets, providing for error recovery, and flow control. X.25 also provides for a number of X.2-defined facilities.

The X.25 DTE-DCE Recommendation actually encompasses the third layer as well as the lower two layers of OSI. Figure 4.5 shows the relationships of the X.25 layers. The recommended physical layer (first layer) in-

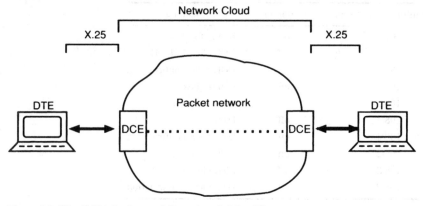

Figure 4.4 The X.25 interface and the network "cloud."

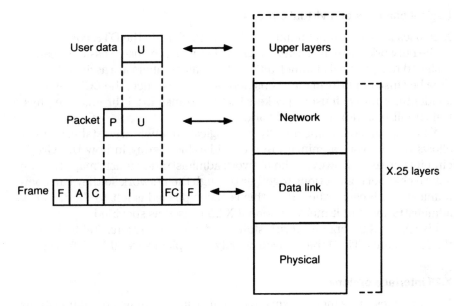

Figure 4.5 The X.25 layers.

terface between the DTE and DCE is X.21. X.25 operates on an X.21 interface by exchanging packets across the T and R interchange circuits when
the interface is in states 13S, 13R, and 13.

X.25 also describes the use of X.21 bis. In order to use this interface, X.25
requires that the following V.24 control circuits be in the on condition: 105,
106, 107, 108, and 109. Packets are then exchanged on the transmit (103)
and receive (104) circuits. If these circuits are off, X.25 assumes the physical layer is in an inactive state and any upper levels (such as data link and
network) will not function. X.25 networks can operate with other physical
layer standards (for example, V.35 and the EIA 232-E standard from the
Electronic Industries Association).

X.25 assumes the data link layer (second level) to be link access protocol,
balanced (LAPB). This line protocol is a subset of high-level data link control (HDLC). It allows the use of LAP, but this older link protocol has largely
fallen in disuse. Some vendors also use other data link controls, such as
Bisync (binary synchronous control), for the link layer.

The X.25 packet is carried within the LAPB frame as the I (information)
field. LAPB ensures that the X.25 packets are transmitted across the link,
after which the frame fields are stripped and the packet is presented to the
network layer. The principal function of the link layer is to deliver the
packet error-free despite the error-prone nature of the communications
link. In X.25, a packet is created at the network level and inserted into a
frame which is created at the data link level.

Logical channels and virtual circuits

X.25 uses logical channel numbers (LCNs) to identify the DTE connections to the network. As many as 4095 logical channels (i.e., user sessions) can be assigned to a physical channel, although not all numbers are assigned at one time because of performance considerations. In essence, the LCN serves as an identifier for each user's packets that are transmitted through the physical circuit to and from the network.

X.25 explains quite specifically how logical channels are established but allows the network administration considerable leeway in how the virtual circuit is created. However, the network administration must "map" the two logical channel connections together through the network so they can communicate with each other. How this is accomplished is left to the network administration, but it must be done if X.25 is used as specified.

Like X.21, X.25 operates with states and state transitions. Table 4.6 lists the X.25 states. This table is used shortly to explain several X.25 features.

X.25 interface options

X.25 provides three mechanisms to establish and maintain communications between the DTEs and the network:

- Permanent virtual circuit (PVC)
- Virtual call (VC)
- Fast select call
- Permanent virtual circuit

TABLE 4.6 X.25 States

State number	State description
p1	Ready
p2	DTE waiting
p3	DCE waiting
p4	Data transfer
p5	Call collision
p6	DTE clear request
p7	DCE clear indication
d1	Flow control ready
d2	DTE reset request
d3	DCE reset indication
r1	Packet layer ready
r2	DTE restart request
r3	DCE restart indication

A PVC is somewhat analogous to a leased line in a telephone network: The transmitting DTE is assured of obtaining a connection to the receiving DTE through the packet network. X.25 requires a PVC to be established before the session begins. Consequently, an agreement must be reached by the two users and the network administration before a permanent virtual connection will be allocated. Among other things, this includes the reservation of an LCN for the PVC user and the establishment of the X.2 facilities.

Thereafter, when a transmitting DTE sends a packet to the network, the LCN in the packet indicates that the requesting DTE has a PVC connection to the receiving DTE. Consequently, a connection will be made by the network and the receiving DTE without further session negotiation. PVC requires no call setup or clearing procedures, and the logical channel is continually in a data transfer state.

Virtual call. A VC (also called a switched virtual call) resembles some of the procedures associated with telephone dial-up lines and the X.21 Recommendation discussed earlier in this chapter. The originating DTE must issue a call request to the network to start the connection operation. In turn, the network (local DCE) relays this packet to the remote DCE, which sends an incoming call packet to the called DTE.

If the receiving DTE chooses to acknowledge and accept the call, it transmits a call accepted packet to the network. The network then transports this packet to the requesting DTE in the form of a call connected packet. The channel enters a data transfer state after the call establishment. To terminate the session, a clear request packet is sent by either DTE. It is received as a clear indication packet and confirmed by the clear confirm packet.

Fast select. Some networks also support a third option, called the fast select. It allows the call connection packets to contain as many as 128 octets of user data. It also allows the clear packets to contain data. The reader should check X.25 for more detail on the fast select.

Example of X.25 operations

Figure 4.6 provides an example of a typical X.25 operation with the VC option. The reader may notice that some of the terms used in the X.21 Recommendation (Figs. 4.2 and 4.3) are also used with X.25. However, X.25 is used to obtain packet-switched services and X.21 is used to obtain circuit-switched services.

As with the X.21, this example shows the operations between the DTE and the DCE in obtaining a connection with the packet network, transferring some data, and releasing the connection. Each event is numbered with an accompanying explanation in the text. Because of the relative simplicity and number of events (unlike X.21), this example is explained with one il-

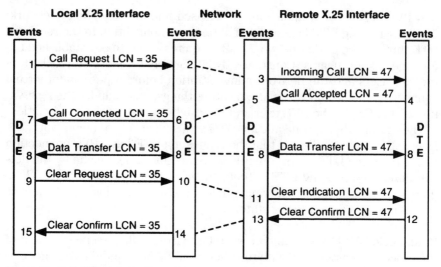

Figure 4.6 Switched virtual call.

lustration. The dashed lines show the relationships of the packets at the network interfaces on both sides of the network. Several (but not all) of the X.25 timers are also brought into this discussion. The reader may find it useful to refer to Table 4.6 during this discussion.

Event 1. The user DTE issues a call request packet to the network (the local DCE). Upon issuing this packet, the DTE enters into the DTE waiting state (state p2) and starts the DTE T21 timer (which is called a DTE time limit). The call request packet must contain the following information:

- The LCN that is to be used at the local interface to identify the call (in this example, LCN=35).

- The called and calling DTE addresses (using X.121 or other addresses, such as an E.164 ISDN address). Some implementations do not require the calling address.

- Requests for facilities, as described in the X.2 Recommendation.

Note: All connection setup and connection clear packets contain all these fields.

Event 2. The network receives the call request packet and makes numerous checks on the validity of the request. X.25 provides many diagnostic codes that the network may use to inform the user about problems with the request (the DTE may also use these codes in its ongoing communications with the network). Assuming all is well, the network makes a routing decision based on the called DTE address and places the packet onto an outgoing trunk in the network.

Event 3. The packet is received at the remote X.25 interface. The network also performs several validity checks on the packet. It determines its destination (on which port the called DTE resides) and selects an available LCN for the connection (LCN = 47). It then presents an incoming call packet to the remote DTE, enters the DCE waiting state (p3) and starts timer T11 (which is called a DCE time-out). During this process, the network is allowed to change the facilities; they may only be downgraded or upgraded to a default value. Also, if the network chooses not to accept the call, it can issue a clear packet to the DTE, with relevant diagnostic information.

Event 4. The remote DTE has the options of accepting the call (as in this example with the call accepted packet) or refusing the call with a clear packet. It can also accept the call by degrading the facilities values or upgrading the values toward a default value. The call accepted packet must contain the same LCN value as the incoming call packet.

Event 5. If the call accepted packet is acceptable to the network, it terminates timer T11, sends the packet to the requesting DCE (local DCE) by the network's routing procedures, and places LCN 47 in the data transfer state (p4).

Event 6. The packet is received from the network and the local DCE sends it to the local DTE as a call connected packet. It must use the same LCN (LCN = 35) as the call request packet.

Event 7. The DTE receives the packet. If it is acceptable (that is, it is well formed, and the facilities that may have been changed by the network or the remote DTE are satisfactory), it turns off its timer T21 and enters LCN 35 into the data transfer state (p4). The machines are now ready to transfer data.

Event 8. Data packets are exchanged between the DTEs through their respective DCEs and the network. X.25 provides many flow control and reliability features during the data transfer process. Some examples follow:

- Use of packet sequence numbers to ensure all traffic is received correctly and properly sequenced

- Use of nondata packets to control the flow of traffic of each LCN session into and out of the network (RR packets for receive ready and RNR packets for receive not ready)

- Segmentation and blocking of data into smaller or larger packets to meet network and DTE needs

- Providing means to reset or restart the connection, depending on particular circumstances, with reset and restart packets, respectively

- Use of Interrupt packets to send unusual or high priority traffic

Event 9. A clear can be initiated by either DTE or the network. In this example, the local DTE sends a clear request packet to the network. In so doing, it enters the DTE clear request state (p6) and starts timer T23.

Event 10. The local DCE sends this packet to the remote DCE.

Event 11. The remote DCE interprets this packet and presents it to the remote DTE as a clear indication packet. It enters LCN 47 into the DCE Clear Indication state (p7) and starts timer T13.

Event 12. The remote DTE responds with a clear confirmation packet.

Event 13. The remote DCE receives this packet and terminates timer T13. It places LCN 47 into the ready state (p1). This figure shows that the clear is transported back to the originating party (see the dashed line). Some networks transport it back, and some do not. If it is not sent back, the clear confirmation can only have local significance, and the originating DCE must take it upon itself to issue the clear confirmation in event 14.

Event 14. Whatever the case, the local DCE issues a clear confirmation packet to the DTE that sent the clear request packet.

Event 15. Upon receiving the clear confirmation, the DTE terminates the T23 timer and enters LCN 35 into the ready state (p1). Hereafter, LCN 35 can be used at the local side to support another call and LCN 47 is available on the remote side as well. Of course, whether these two LCNs are once again mapped together to form an end-to-end virtual circuit is simply a matter of the local DTE and the remote DCE choosing the same values from a pool of LCN numbers.

Comparison of 1980, 1984, and 1988 X.25 releases

The 1980 Recommendation of X.25 was the first release of X.25 that achieved worldwide use. However, today, the vast majority of systems now use the Red Book 1984 or Blue Book 1988 Recommendations. The 1984 release contains a substantial number of changes to the 1980 version. The most significant changes are summarized in Table 4.7.

TABLE 4.7   Major Changes between the X.25 1980 and 1984 Recommendations

- The datagram is deleted.
- The fast select facility is now essential, and user data can reside in both call setups and clearings.
- The interrupt data packet user data field is increased to 32 octets.
- Facility fields in the call management packets are increased to 109 octets.
- The transit delay facility allows the negotiation of delay: it is added as an essential facility.
- Additional error and diagnostics codes are added.

TABLE 4.7   (Continued)

- Several telephone and PBX type services are added:
  - ~ Call redirection notification
  - ~ Hunt groups
  - ~ Local charging prevention
  - ~ Network user Identification
  - ~ Called line address modification
- The recognized private operating agencies (RPOA) facility is added to allow the selection of transit networks.
- The registration facility is added, and timer (time limit) T28 is added to control registration facility timer operations.
- Several end-to-end services are added:
  - ~ Address extension
  - ~ Interrupt procedures
  - ~ Minimum throughput class negotiation
  - ~ End-to-end transit delay negotiation
- A DTE can now belong to up to 10,000 closed user groups.
- Services are added to facilitate the interconnection of public and private networks.
- Several additions are made at the link level:
  - ~ Extended sequencing (modulo 128)
  - ~ Multilink procedures (MLP)
  - ~ An idle timer (T3)

The 1988 Recommendation that is published in the ITU-T Blue Books also has some changes, but they are not as significant as the 1984 changes. The major changes reflected in the 1988 Recommendation are listed in Table 4.8.

The PAD "Triple X" Recommendations

As the packet networks began to flourish in the 1970s, the standards groups, such as the ITU-T, recognized that the majority of terminals in operation were (and are) asynchronous devices. Obviously, an interface was needed to connect these terminals into packet networks. Consequently, standards were developed to provide *protocol conversion* and *PAD* functions for the asynchronous terminal. After the initial 1976 draft of the X.25

TABLE 4.8   Major Changes between the 1984 and the 1988 X.25 Recommendations.

- Notations on the use of X.31 have been added to the physical layer.
- Transit delay definitions have been made more explicit.
- Discussions on throughput considerations have been amplified.
- Section 4.6, physical and data link failures, has been expanded.
- The long address (A bit) format and facility have been added.
- The call deflection facility has been added.

standard, the standards committees followed up in 1977 with recommendations for three specifications to support X.25 with asynchronous terminal interfaces: *X.3, X.28,* and *X.29.* These recommendations have been enhanced with the 1988 release. This chapter provides a general description of X.28 and X.29. X.3 is covered in chapter 3.

X.28

This standard defines the procedures to control the data flow between the user terminal and the PAD (see Figure 4.7). Upon receipt of an initial transmission from the user DTE, the PAD establishes a connection and provides services according to X.28. The user DTE evokes X.28 commands to the PAD, which then requests an X.25 virtual call to a remote DTE.

The PAD is responsible for transmitting the appropriate X.25 call request packet. Messages sent from the terminal to the PAD are called *command signals* and messages sent from the PAD to the terminal are called *service signals.* X.28 supports the procedures for:

The establishment of the path

The initialization of service

The exchange of data

The exchange of control information

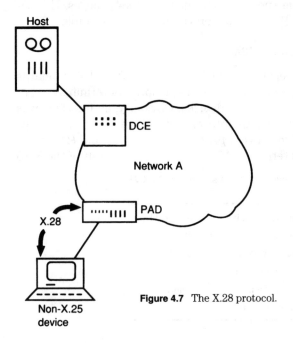

Figure 4.7 The X.28 protocol.

X.28 requires the PAD to return a response when a terminal issues a command to it. It also specifies that two profiles can be defined for providing service to the user DTE. The *transparent* profile means the servicing PAD is transparent to both DTEs: The DTEs perceive that they have a direct virtual connection to each other. In this situation, the remote DTE is responsible for some PAD functions such as error checking. The *simple* profile makes use of the fully defined X.3 Standard and the parameter functions to satisfy the user DTE requests. The 1988 version of X.3 provides a user with the flexibility to tailor additional characteristics for a particular terminal.

X.28 physical layer requirements. X.28 defines the use of the ITU-T V Series or X Series modems for the physical layer interface. Lower-speed interfaces use the V.21, V.22, and V.23 standards (300 to 1200 bit/s). Medium-speed interfaces use V.22 bis (2400 bit/s). V.28 is stipulated for use, although this is an unnecessary requirement because the V Series interfaces cite this use of V.28 within their own respective standards. Automatic dial-in and dial-out uses V.25, although in North America, the EIA 366-A is more prevalent. X.28 does not cite the use of any other higher-speed V Series interfaces for the physical layer.

Call establishment between the DTE and the PAD. X.28 describes the DTE and PAD call establishment procedures through state logic and state diagrams. This chapter provides a brief overview of the call establishment state logic. Like many other ITU-T Recommendations, X.28 makes use of state transition diagrams, and this section takes the reader through one of them with a verbal description.

After the physical path between the DTE and DCE has been established, the DTE and the PAD exchange binary 1s between the DTE-DCE interface. The interface is then in an active link state. At this time the DTE is permitted to transmit characters indicating a service request, in order to initialize the PAD.

The PAD is configured to be an auto-baud port. This means that the PAD is able to automatically detect the data rate, code, and parity used by the DTE. In the past, auto-baud meant automatic detection of bit rate, but its definition has been expanded.

After the auto-baud operations, the PAD selects the initial profile for the terminal. The PAD may choose standard profiles for the X.3 parameters. These standard profiles are described in X.28.

Certain actions may be unnecessary if the terminal is connected to the PAD directly or connected through a leased line. For example the PAD may already know the speed, code, and initial profile of the terminal. Obviously in this case, the previous operations are unnecessary. In any event, following the transmission of the service request signal, the DTE enters the DTE wait state (which is defined as state 3A). This wait state is actually stipulated if the value of parameter 6 is set to zero. If this parameter is not set to zero, the interface enters a service ready state directly upon the PAD transmitting a PAD identification service signal. This signal is sent after the PAD receives a service request signal from the DTE.

After the transmission of the PAD service signal, the interface enters a PAD waiting state unless a virtual call is established or is being established. Even when the PAD is in the PAD waiting state (state 5), the DTE is permitted to transmit a PAD command signal. Following the transmission of the PAD command signal, the DTE must enter into the DTE wait state.

If parameter 6 is not set to zero, the PAD upon receiving a valid PAD selection command signal will transmit an acknowledgment PAD service signal and will enter a connection in progress state (state 7). During this period the PAD is not allowed to accept any PAD command signals. The transition to state 8 (service signal state) occurs when the PAD receives a call destined for the DTE. The PAD service signal state allows a transition to data transfer, connection in progress, or PAD waiting. Other scenarios allow the PAD and the terminal to move from the data transfer state back to the service signal state or the waiting for command state.

X.29

This standard provides the procedures for the PAD and a remote DTE or PAD to exchange control information on an X.25 call. X.29 is quite useful in allowing a host computer (DTE) to change the PAD parameters of terminals connected to it. X.29 allows the exchange of information to occur at any time, either at a data transfer phase or any other phase of the virtual call. Figure 4.8 shows the relationship of X.29 to the DTEs and the DCEs.

The X.25 Q bit controls certain functions of X.29. An X.25 packet contains user data fields and may also contain the Q bit. The Q bit (or data qualified bit) is contained in the header of the data packet. It is used by the remote DTE to distinguish between a packet containing user information (Q = 0) and one containing PAD information (Q = 1).

X.29 defines seven control messages, which are called PAD messages. These messages are:

Set	Changes an X.3 value.
Read	Reads an X.3 value.
Set and Read	Changes an X.3 value and requires the PAD to confirm the change.
Parameters Indication	Returned in response to the above commands.
Invitation to Clear	Allows X.25 Call Clear by a remote DTE. The PAD clears to the local terminal.

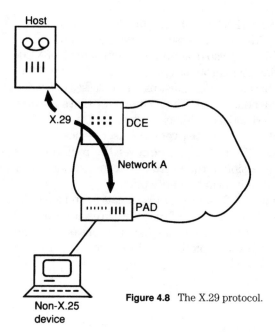

Figure 4.8 The X.29 protocol.

Indication of Break	PAD indicates that the terminal has transmitted a break.
Error	The device reacts in response to an invalid PAD message.

The X.29 protocol can be used to change a terminal's profile. In many instances, the terminal user (an accountant, an automated teller machine customer, etc.) has no idea about the network and its many PAD parameters. The X.29 feature gives a host computer the capability to reconfigure certain characteristics of the devices that communicate with it through a PAD. Of course, how the PAD facilities are used is largely up to the creativity and imagination of the network managers and designers.

X.30

The ITU-T recognizes that the X Series interface recommendations will be prevalent for many years. From the perspective of ITU-T, it is desirable to evolve quickly to all digital systems using the ISDN interface standards. From the perspective of the end users, this may not be quite as desirable. Nonetheless, X.30 was published by the ITU-T to aid in the transition from the X Series interfaces to an all digital system using the ISDN standards (I and Q Series).

X.30 describes the connections of X.21 and X.21 bis devices to an ISDN. These devices must operate within the user classes of service 3 to 7 and 19, according to Recommendation X.1. X.30 also describes the connections of X.20-based devices with an ISDN using asynchronous data rates of 600, 1200, 2400, 4800, and 9600 bit/s. The X.20 bis connection must use user classes of service 1 and 2 as stipulated in Recommendation X.1. The recommendation stipulates the use with both circuit-switched and leased-line systems.

X.30 covers the rate adaptation scheme between the user device through the user ISDN terminal adapter (TA). It does not cover the requirements for the data transfer speed conversion in the event of internetworking—for example between ISDNs and circuit-switched networks.

X.30 also defines the mapping of X.21 and X.21 bis signals to and from the ISDN network layer messages. Two examples of these features are provided in Figures 4.9 and 4.10. The reader can refer to the explanations of X.21 and X.21 bis in this chapter for a more complete description of the X.21 and X.21 bis operations.

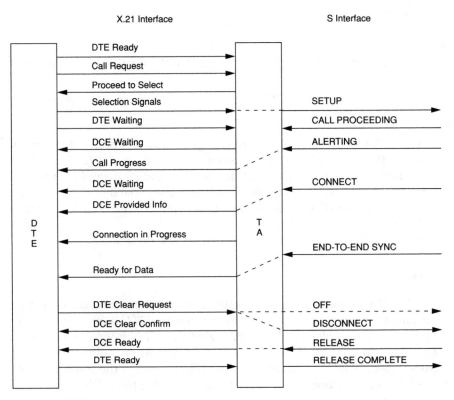

Figure 4.9 X.30 operations.

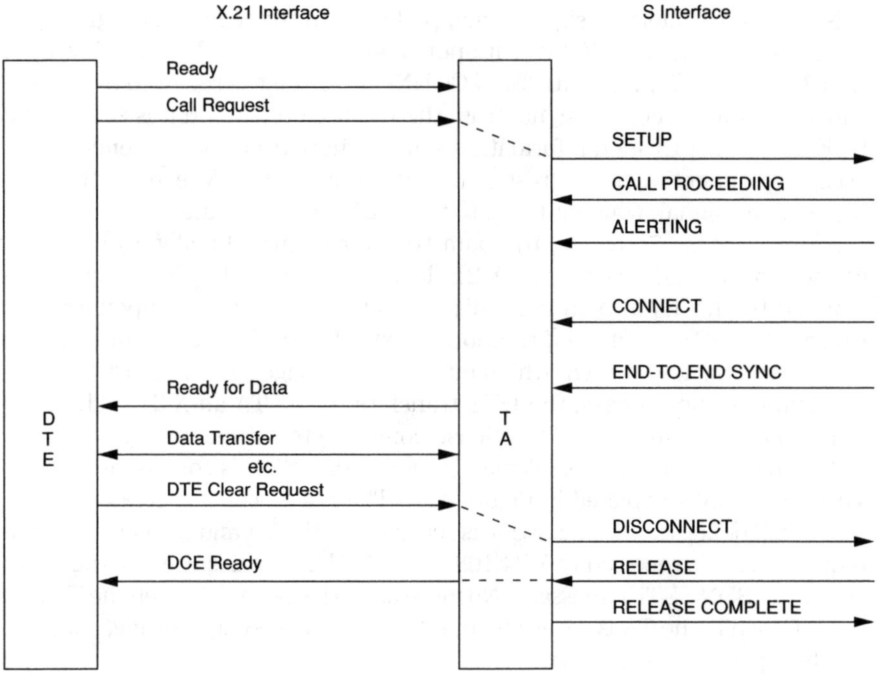

X.21 Interface   S Interface

Figure 4.10 X.30 operations.

Figure 4.9 shows the X.21 mapping functions; Figure 4.10 shows the X.21 bis mapping functions. The X.21 operations begin with the DTE issuing the DTE ready signal to the TA. The X.21 DTE ready signal is conveyed across the X.24 interface with circuits t and c in the following states: t = 1 and circuit c = off. The TA receives the DTE ready signal and then waits for the DTE call request. As seen in Figure 4.9, this occurs with the DTE transmitting the call request signal. The call request signal is conveyed by the following signal across the X.24 circuits: t = 0, c = on. (Hereafter in our description of this operation, we are not going to describe the on/off and 0/1 relationships of the X.24 interfaces. The reader may refer to X.21 and Figures 4.2 and 4.3 for a description of these signals.)

Continuing our analysis, the TA receives the call request signal and returns a proceed to select signal. In turn, the DTE sends the selection signals to the TA.

By this time, the TA has enough information to map the selection signals into an ISDN call establishment with an ISDN I.450/451 SETUP message. The DTE then informs the TA it is in a DTE waiting state. During this time, the TA receives status messages across the S/T reference point. The status messages are shown as call proceeding and alerting in Figure 4.9.

Note that the alerting signal is mapped back to the X.21 call progress signal. Note also that the TA takes it upon itself to return an X.21 DCE waiting signal to the DTE. Finally, an ISDN CONNect message is received by the TA, which is a Call Accepted signal from the remote entity, and it is sent to the DTE in a DCE-provided information signal. Thereafter, the TA sends to the DTE the connection in progress signal. Eventually, the TA receives the end-to-end sync signal, which it maps to the X.21 ready for data signal.

After all these activities, the data transfer occurs. Finally, to clear the connection, the DTE issues the X.21 clear request signals. These signals are mapped by the TA into an ISDN disconnect. The TA takes it upon itself to return the DCE clear confirmation. Next, the ISDN release message is transferred back to the TA, which maps this message to an X.21 DCE ready. To complete the process, the DTE transfers to the TA an X.21 DTE ready signal. The TA issues an ISDN release complete to its peer entity.

The mapping of X.21 bis interface to and from ISDN is somewhat simpler. This operation is depicted in Figure 4.10. The ready signal is created by the V.24 108.2 being set off. This signal is received by the TA and held until the call request signal is received (a V.24 108/1 = on). This signal is used by the TA to create an ISDN SETUP message. No more activities occur between the TA and the DTE while the TA is receiving an ISDN call proceeding, alerting, connect, and finally an end-to-end sync.

Upon receiving this last message, the TA turns on the ready-for-data signal to its DTE (V.24 107 = on). Upon receiving the signal, the DTE enters the data transfer mode and transfers data. Eventually the DTE clears the connection by a DTE clear request (V.24 108/1 = off). This signal begins the ISDN disconnect operations depicted in the bottom of Figure 4.10.

X.30 mapping and framing operations

A system using X.30 must go through some rather elaborate operations to interface with an ISDN. These operations are accomplished through a TA. An X.30 TA is shown in Figure 4.11. It is quite similar to TAs found in other ISDN interworking operations (for example, V.110 has a similar TA). The TA consists of two functions, rate adaptation 1 (RA1) and rate adaptation 2 (RA2). The purpose of RA1 is to convert the X.21 or X.21 bis data rate to an intermediate rate (IR) of either 8 or 16 kbit/s. It uses ITU-T Recommendation X.1 to define the classes of service and the data signalling rate.

X.1 rates for user classes 3, 4, and 5 are converted to a 8-kbit/s rate and user class 6 is converted to a 16-kbit/s rate. The intermediate rate is then input into RA2, which is responsible for placing the data onto the ISDN 64-kbit/s basic channel.

The output of the RA1 function is in the form of a 40-bit frame. The structure of this frame is shown in Figure 4.12. This frame is similar to the frame used in V.110 TAs. The frame, although described in 40 bits, actually con-

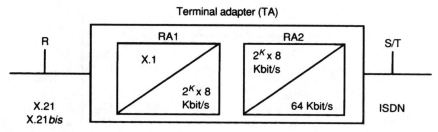

Figure 4.11 The X.30 TA.

sists of two multiframes. The odd-numbered frames set the octet position 1 to all 0s. The even-numbered frames use this octet for the E bits (which are described later).

Figure 4.13 shows the position of the data bits in the odd and even frames (note the boldface areas). The data bits are placed in these positions, low-order bit first, in three octets in each of the frames. Therefore, each frame carries 24 data bits in the P, Q, and R bit positions.

Figure 4.14 shows the status bits. They are labeled SQ, SR, SP, and X. The status bits provide a mapping from the local X.21 interface to the remote X.21 interface. We shall shortly see that the mapping occurs by the X.30 protocol establishing a relationship between the SP, SQ, SR, and the data bit groups of P, Q, and R. For our present analysis, it is sufficient to note that the S bits are used on the transmit side (local side) to reflect the state of the X.21 c interchange circuit. These bits are then conveyed across an ISDN node to the remote receiver, where they are used to create the signals on

				Bit position					
1	2	3	4	5	6	7	8	Octet position	
0	0	0	0	0	0	0	0	0	
1	P1	P2	P3	P4	P5	P6	SQ	1	Odd
1	P7	P8	Q1	Q2	Q3	Q4	X	2	frames
1	Q5	Q6	Q7	Q8	R1	R2	SR	3	
1	R3	R4	R5	R6	R7	R8	SP	4	
1	E1	E2	E3	E4	E5	E6	E7	0	
1	P1	P2	P3	P4	P5	P6	SQ	1	Even
1	P7	P8	Q1	Q2	Q3	Q4	X	2	frames
1	Q5	Q6	Q7	Q8	R1	R2	SR	3	
1	R3	R4	R5	R6	R7	R8	SP	4	

Figure 4.12 The X.30 frame.

Bit position								Octet position
1	2	3	4	5	6	7	8	
0	0	0	0	0	0	0	0	0
1	P1	P2	P3	P4	P5	P6	SQ	1 Odd
1	P7	P8	Q1	Q2	Q3	Q4	X	2 frames
1	Q5	Q6	Q7	Q8	R1	R2	SR	3
1	R3	R4	R5	R6	R7	R8	SP	4
1	E1	E2	E3	E4	E5	E6	E7	0
1	P1	P2	P3	P4	P5	P6	SQ	1 Even
1	P7	P8	Q1	Q2	Q3	Q4	X	2 frames
1	Q5	Q6	Q7	Q8	R1	R2	SR	3
1	R3	R4	R5	R6	R7	R8	SP	4

Figure 4.13 Data in the X.30 frame.

the X.21 i interchange circuit. (In all cases during this discussion the c and i circuits represent an on condition with binary 0 and off condition with binary 1. The contents of the SQ, SR, and SP bits are: 0 = on, 1 = off.)

Figure 4.15 shows the relationships of the X.24 circuits (at the transmit and receive side) and the SQ, SR, and SP bits. Notice that the on and off signals at the transmit side c are mapped into the SQ, SR, and SP bits, which are then used to set the proper signal on the receive i interchange circuit.

Figure 4.16 shows the mapping of the transmit c interchange circuit to the receive i interchange circuit. The timing for the mapping occurs

Bit position								Octet position
1	2	3	4	5	6	7	8	
0	0	0	0	0	0	0	0	0
1	P1	P2	P3	P4	P5	P6	SQ	1 Odd
1	P7	P8	Q1	Q2	Q3	Q4	X	2 frames
1	Q5	Q6	Q7	Q8	R1	R2	SR	3
1	R3	R4	R5	R6	R7	R8	SP	4
1	E1	E2	E3	E4	E5	E6	E7	0
1	P1	P2	P3	P4	P5	P6	SQ	1 Even
1	P7	P8	Q1	Q2	Q3	Q4	X	2 frames
1	Q5	Q6	Q7	Q8	R1	R2	SR	3
1	R3	R4	R5	R6	R7	R8	SP	4

Figure 4.14 The status bits. X.

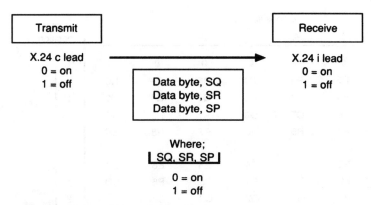

Figure 4.15 The c and i interchange signals.

through the eighth bit position of the R, P, and Q data octets. The arrows in Figure 4.16 show that the sampling of the c lead occurs in the middle of the eighth bit of the respective preceding R, P, or Q bit group. At the receive side, the SQ, SP, and SR status bits are used to create the signals on the i interchange circuit. In essence, on the receive side, the values of status bits SP, SQ, and SR are adapted by the i lead.

Figure 4.17 illustrates the frame synchronization bits. X.30 uses a 17-bit frame alignment pattern with eight 0s in the first octet of the odd frame and the first bit positions of subsequent octets in the odd and even frames. These bit values are set to 1.

The E bits in Figure 4.18 are set to various values to indicate the signalling rate. Their values can be set to indicate a signalling rate of 600, 2400, 4800, or 9600 bit/s. The reader should be aware that the E bits in X.30 are not used in exactly the same way as they are used in V.110. For example, bits E4, E5, and E6 may be used in Recommendation V.110 for network clocking information.

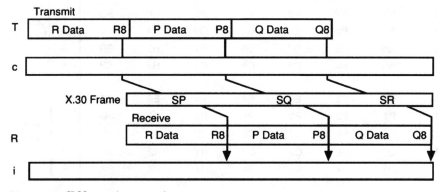

Figure 4.16 X.30 mapping operations.

Bit position								Octet position	
1	2	3	4	5	6	7	8		
0	0	0	0	0	0	0	0	0	
1	P1	P2	P3	P4	P5	P6	SQ	1	Odd frames
1	P7	P8	Q1	Q2	Q3	Q4	X	2	Odd frames
1	Q5	Q6	Q7	Q8	R1	R2	SR	3	
1	R3	R4	R5	R6	R7	R8	SP	4	
1	E1	E2	E3	E4	E5	E6	E7	0	Even frames
1	P1	P2	P3	P4	P5	P6	SQ	1	Even frames
1	P7	P8	Q1	Q2	Q3	Q4	X	2	
1	Q5	Q6	Q7	Q8	R1	R2	SR	3	
1	R3	R4	R5	R6	R7	R8	SP	4	

Figure 4.17 The frame synchronization bits.

Figures 4.19 and 4.20 show two examples of how the data rates are mapped into the X.30 frame. Figure 4.19 shows the frame used to adapt a 2400-bit/s user rate to a 8-kbit/s bearer rate. Notice that user bit 1 is coded into the frame twice as P1, P1. The same scheme applies to the other bits: P2, P2, P3, P3, etc. A simple calculation will show how these bits are used:

$$(8000 \text{ bit/s})/80 \text{ bits/frame}) = 100 \text{ frames/s}$$
$$100 \text{ frames/s} \times 48 \text{ user bits/multiframe} = 4800 \text{ bit/s}$$

Every other data bit position contains a data bit; therefore the data rate is 2400 bit/s.

Bit position								Octet position	
1	2	3	4	5	6	7	8		
0	0	0	0	0	0	0	0	0	
1	P1	P2	P3	P4	P5	P6	SQ	1	Odd frames
1	P7	P8	Q1	Q2	Q3	Q4	X	2	Odd frames
1	Q5	Q6	Q7	Q8	R1	R2	SR	3	
1	R3	R4	R5	R6	R7	R8	SP	4	
1	**E1**	**E2**	**E3**	**E4**	**E5**	**E6**	**E7**	0	Even frames
1	P1	P2	P3	P4	P5	P6	SQ	1	Even frames
1	P7	P8	Q1	Q2	Q3	Q4	X	2	
1	Q5	Q6	Q7	Q8	R1	R2	SR	3	
1	R3	R4	R5	R6	R7	R8	SP	4	

Figure 4.18 The E bits.

1	2	3	4	Bit position 5	6	7	8	Octet position	
0	0	0	0	0	0	0	0	0	
1	P1	P1	P2	P2	P3	P3	SQ	1	Odd frames
1	P4	P4	P5	P5	P6	P6	X	2	
1	P7	P7	P8	P8	Q1	Q1	SR	3	
1	Q2	Q2	Q3	Q3	Q4	Q4	SP	4	
1	1	1	0	E4	E5	E6	E7	0	
1	Q5	Q5	Q6	Q6	Q7	Q7	SQ	1	Even frames
1	Q8	Q8	R1	R1	R2	R2	X	2	
1	R3	R3	R4	R4	R5	R5	SR	3	
1	R6	R6	R7	R7	R8	R8	SP	4	

Figure 4.19 The 2400 bit/s frame.

There are excessive data bit slots relative to the 2400-bit/s data rate. Therefore, not all the slots are used. As shown in Figure 4.19, every other slot is really a redundant copy of the previous data bit.

The adaptation of a 4800-bit/s user rate to the 8-kbit/s bearer rate is more straightforward. Figure 4.20 shows this relationship, which can be calculated as follows (which is the same as the 2400 bit/s rate, but without skipping every other bit position):

$$(8000 \text{ bit/s})/80 \text{ bits/frame}) = 100 \text{ frames/s}$$
$$100 \text{ frames/s} \times 48 \text{ user bits/multiframe} = 4800 \text{ bit/s}$$

1	2	3	4	Bit position 5	6	7	8	Octet position	
0	0	0	0	0	0	0	0	0	
1	P1	P2	P3	P4	P5	P6	SQ	1	Odd frames
1	P7	P8	Q1	Q2	Q3	Q4	X	2	
1	Q5	Q6	Q7	Q8	R1	R2	SR	3	
1	R3	R4	R5	R6	R7	R8	SP	4	
1	0	1	1	E4	E5	E6	E7	0	
1	P1	P2	P3	P4	P5	P6	SQ	1	Even frames
1	P7	P8	Q1	Q2	Q3	Q4	X	2	
1	Q5	Q6	Q7	Q8	R1	R2	SR	3	
1	R3	R4	R5	R6	R7	R8	SP	4	

Figure 4.20 The 4800 bit/s frame.

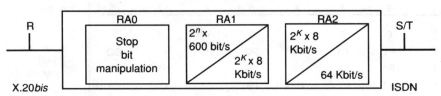

Figure 4.21 The X.30 RA0 TA.

X.30 also supports terminal adaptation operations from terminals using X.1 user classes of service 1 and 2. The reader may recall from previous discussions of X.1 that user classes of service 1 and 2 define asynchronous operations. The adaptation is provided through yet another TA as shown in Figure 4.21. This TA has another stage labeled RA0 (rate adapter 0). This stage functions as an asynchronous-to-synchronous convertor and uses the conversion techniques defined in ITU-T V.14.

X.31

The ITU-T Recommendation X.31 provides two scenarios (cases) for the interface of a X.25 packet-mode terminal into an ISDN node: case A access to a packet network and case B. This discussion assumes the reader has a basic level of understanding of ISDN terminology. Appendix B provides a tutorial on ISDN.

Case A—Access to a packet data network

Figure 4.22 illustrates case A. It supports a rudimentary and basic service. The ISDN provides a transparent handling of the packet calls from the DTE to the X.25 network. This scenario only supports B channel access. Moreover, if two local DTEs wish to communicate with each other, their packets must be trans-

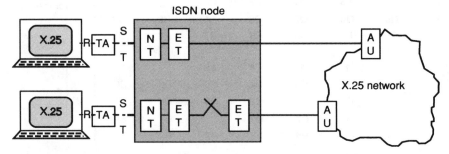

Figure 4.22 Case A with X.31; minimum configuration scenario.

mitted through the ISDN and through the X.25 network before the packets can be relayed to the other DTE. This minimum integration scenario uses the B channels for all call management. This is performed using ISDN signalling procedures prior to initiating the X.25 level 2 and level 3 procedures. In essence, the ISDN node passes the X.25 call transparently to the X.25 network.

Figure 4.22 shows two options that can be obtained with case A. Please note that in this figure (and in all ISDN specifications), an X.25 and its TA are considered equivalent to a packet-mode TE1 at the ISDN S/T interface. The top part of the figure illustrates how the permanent, nondemand option is used. The X.25 DTE and TA or the TE1 is connected to the ISDN port at the packet data network ISDN access unit (AU) port. The ISDN X.931 messages are not used, and the TA is responsible for physical rate adaptation (bit rate changes) between the R reference point and the 64-kbit/s B channel rate.

The bottom part of Figure 4.22 shows the second option, called *demand access*. Prior to any X.25 functions, ISDN signalling procedures are used to set up a B channel. Typically, an ISDN Q.931 SETUP message identifies the access port of the packet data network, and the called DTE field in the X.25 packet is used to identify the remote user device.

Case B—Using the ISDN virtual circuit service

The case B scenario provides several additional functions, as shown in Figure 4.23. The ISDN provides a packet handler (PH) function within the node. Actual implementations of case B use two separate facilities for

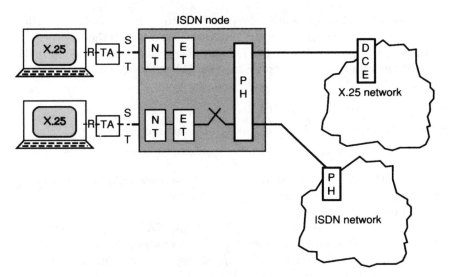

Figure 4.23 Case B with X.31; maximum configuration scenario.

the interconnection. The ISDN switch is provided through a vendor's ISDN NT1 product and the PH handler function is provided by the vendor's packet switch.

Case B permits two options: (1) access via the B channel and (2) access via the D channel. With the first option, the X.25 packet and data link layer procedures are conveyed through the B channel. The access through the D channel requires that all active logical channels are established through a D channel connection and all X.25 packets, including connection setups, connection disconnects, and data packets, must be transmitted on the D channel on the LAPD link.

Example of X.25 and ISDN Q.931 operations

Figure 4.24 shows one example (several scenarios are possible) of how X.31 defines the mapping of the X.25 protocol and IDSN messages. The figure is almost self-explanatory, although the reader may need to refer to appendixes A and B and the X.25 material for tutorials on data link controls and ISDNs, respectively. The top "boxes" represent the components that participate in the operation:

- TE2 A and TE2 B: The originating and receiving user devices, respectively
- TA1 and TA2: The originating and receiving TAs, respectively
- NT1: The ISDN physical connections between the TA and the ISDN node
- LE: The ISDN local exchanges
- PH/NW: The ISDN PH and/or the packet network (NW)

The top left part of the figure shows the TA setting up an LAPD connection with the LE with SABME and UA. This illustration assumes the X.21 or X.21 bis operations between TE1 A and TA1 have been completed. Therefore, the physical layer interface between TE2 A and TA1 (and the remote TE and TA as well) is:

- For X.21: States 13S, 13R, or 13
- For X.21 bis: V.24 circuits 105 through 109 in on condition
- For EIA-232-E: Circuits CA, CB, CC, CD, and CF in on condition

Next, a B channel is created with Q.931 messages (the SETUP, CALL PROceeding, CONNect, and CONNect ACK messages) between the TA, the LE, and the PH or the NW.

After the creation of a B channel, TE2 A and TA1 set up an LAPB connection with SABM and UA. Then, the TE2 A enters into the X.25 packet level procedures by sending a Call Request packet to the TA. This packet is relayed to the LE and then to the PH/NW.

Figure 4.24 Mapping ISDN and X.25.

The next series of operations on the receiving side simply mirror the operations that we have examined on the originating side. The only difference is the changing of the X.25 Call Request packet to the X.25 Incoming Call packet.

The X.25 Call Accepted packet is returned from the called DTE and is mapped to the Call Confirmation packet for transfer to the originating DTE.

After these operations, ISDN becomes a transparent pipeline for the exchange of X.25 packets.

The remainder of the operations in this figure show the disconnection operations in the following order:

- The release of the X.25 virtual circuit (Clear Request, Clear Indication, and Clear Confirmation packets)
- The release of the LAPB connection (the DISC and UA operations)
- The release of the B channel (the Q.931 RELease, and RELease COMplete messages)
- The release of the LAPD connection (the DISC and UA operations)

In an actual situation, the LAPD connection and the B channel may not be taken down. Moreover, the LAPB connection may also remain "nailed up."

Figure 4.25 shows the use of the D and B channels involved in these operations. The lighter shade illustrates the ISDN activity and the darker shade illustrates the X.25 activity. X.25 data and control packets may be transmitted in the B channel slots. Many implementations of X.31 also permit these packets to be sent in the D channel.

X.32

X.32 is published by the ITU-T to provide guidelines for the interfaces of user devices (DTE) and DCEs through (1) a switched telephone network, (2) an ISDN, (3) a packet-switched network, or (4) a circuit-switched data network to communicate with a packet-switched network. The ITU-T prefers the use of public networks with X.32, but private networks can use X.32 equally well.

X.32 Functions

The functional view of X.32 is provided in Figure 4.26. The operations shown in this figure are called dial-in by the DTE and dial-out by the public-switched packet data network (PSPDN). The dial-in operation is used by the packet-mode DTE to access the data network by means of X.32-based selection procedures. The DTE uses automatic or manual dial-up ports and modems. The dial-out from the packet network to the user device uses automatic answering procedures. Optionally, the DTE can use manual answering.

When communications are to be established on dial-up, switched links, it is often a good idea to require identification procedures between the calling and called parties. X.32 stipulates procedures for the establishment of both a DTE and DCE *identity*. When a DTE dials in to a network or is accessed by the network, there may be a requirement by the network for the identification of the DTE to the DCE. Likewise, the DTE may wish to receive the identity of the DCE before proceeding with the exchange of data.

DTE services

X.32 defines three types of DTE services: (1) unidentified DTE service, (2) identified DTE service, and (3) customized DTE service (see X.2 for a definition of customized DTE service).

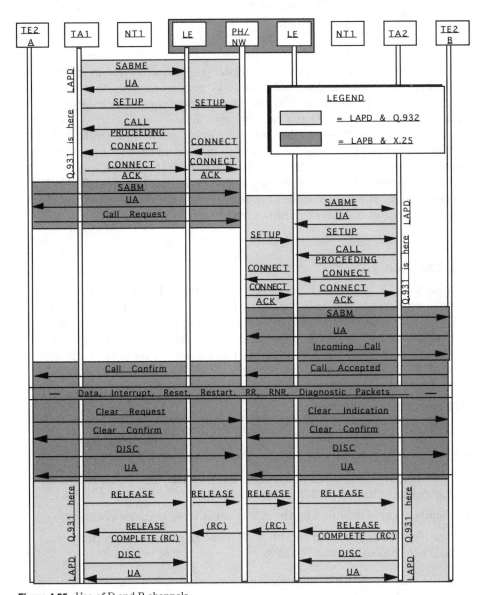

Figure 4.25 Use of D and B channels.

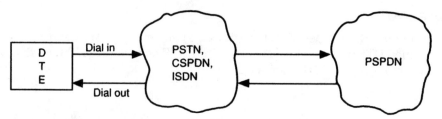

Figure 4.26 X.32 functional view.

The unidentified DTE service is available as both DTE dial-in and network dial-out. The DTE does not have to be registered with the network. However, it may not be allowed to make paid calls or receive reverse charge calls. This possible restriction is network specific.

The identified DTE service has the same effect as the unidentified DTE service except the DTE may be allowed to make paid calls and receive reverse charge calls. In effect, the DTE is billable with this option.

The customized DTE service has the same effect as the identified DTE service except the DTE's address is registered with the network and its X.2 facilities and link layer parameters are tailored to the DTE requirements.

The X.32 layers

The layers of X.32 are shown in Figure 4.27. Refer to this figure as each layer is examined. The half-duplex management operations are optional and may not be implemented in full-duplex systems. All other layers are required. This figure does not show the protocol data units (PDUs) used in the layers. The reader may wish to refer to Figure 4.5 for a more detailed view of the PDUs (frames and packets).

DTE and DCE identification

X.32 provides four options for DTE identification:

- Identification is provided by the public-switched network.
- Identification is provided at the link layer by the use of the HDLC exchange identification (XID) frame.
- Identification is provided by the X.25 packet layer registration facility.
- Identification is provided by the X.25 network user identification (NUI) facility in the Call Request packet.

In the case of the DCE identification procedures, X.32 stipulates three choices for this procedure:

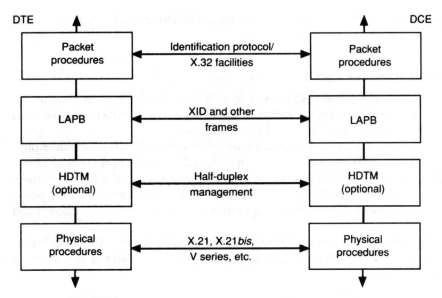

Figure 4.27 The X.32 layers.

- The identification is provided by the public-switched network.
- Identification is provided by the link level XID frame.
- The identification is provided by packet layer registration procedure called the identification protocol (discussed later).

In both the DTE and DCE identification procedures, X.32 permits the identity of these devices to become known with one of two procedures. The first procedure allows the identification to be established prior to any virtual call establishment (i.e., before any X.25 call management packet can be exchanged between the DTE and DCE, the identification must be completed). The second choice allows the identification to be performed per virtual call by means of the NUI facility. With this method, the identification of the DTE is coded in the facility field within the Call Request packet.

In establishing the DCE identity, X.32 stipulates that the identification (prior to virtual call establishment) can be provided by the use of a public-switched network through the ISDN number or the circuit-switched public data network number that identifies the DCE. As we just mentioned, the DCE identity procedure can also be provided by the HDLC XID frame procedure.

The T14 timer

The network may use the T14 timer to control the identification operation. It provides a means to assure that the process is completed within a reasonable time and terminated if the DTE does not follow the identification procedures within the T14 period. Figure 4.28 illustrates how the T14 could be employed by a network.

Physical and link layer operations

The physical layer for X32 operations is X.21, X.21 bis, or several of the ITU-T V Series modems. Practically speaking, other interfaces may be used by networks, such as RS-422, T1 signalling, etc.

The data link layer can be either one of two of the HDLC procedures published by the ISO: (1) Class BA with 2, 8, or 2, 8 and 10 options or (2) class BA with 1, 2, 8 or 1, 2, 8, and 10 options. The former is the LAPB protocol and the latter provides the XID for DTE-DCE identification and for selection of optional X.32 facilities. Appendix A explains the HDLC classes and options.

The address field of the HDLC frame is used to (1) identify the called or calling party and (2) determine if the frame is a command or response. Its contents are:

	Calling A	Called B
Command	B	A
Response	A	B

The information (I) field in the HDLC frame consists of the following subfields:

- Format identifier subfield: An ISO field of eight bits = 01000001
- Layer subfield: Not defined at this time
- User data subfield: Contains the X.32 identification protocol code or the X.32 facilities

The HDTM. The HDLC operations can be extended to support half-duplex transmission by using the half-duplex transmission module (HDTM). Figure 4.27 shows the structure of the X.32 link layer if it includes HDTM. The HDTM modules in the DTE and DCE take turns in sending data in the following manner: The calling machine has the initial right to transmit. After the machine is through, it sends an idle signal (at least 15 binary 1s). This signal turns the line around and allows the other machine to send. It must then send at least five flags (0111110) to acknowledge the line turn-around. If the flags are not sent, the relinquishing DTE-DCE may assume transmission rights by sending flags.

An option for half-duplex operations is to use the V.24/EIA 232-E carrier off signal to detect an idle condition and the relinquishing of the line.

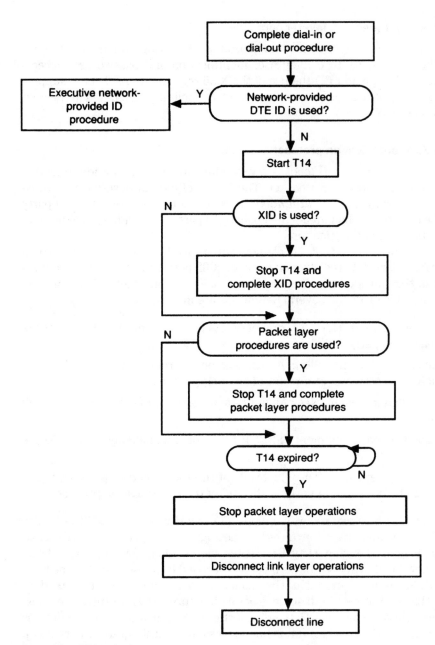

Figure 4.28 T14 operations.

X.32 does not define how the HDTM module inhibits LAPB from sending data in its normal full-duplex mode. However, it does provide some guidelines in Appendix 1 of the recommendation.

Packet layer (network layer) operations

The X.32 packet layer operations permit the following options: (1) the use of the NUI facility to convey user identification and passwords in accordance with X.25 and (2) the use of the X.25 registration packets to support the X.32 identification protocol and the X.32 facilities (discussed in the next two sections).

The X.32 identification protocol

The *identification protocol* is a procedure for exchanging identification and authentication information. The X.32 definitions describe the *questioning* party and the *challenged party*. Obviously, the questioning party is the one that is challenging the challenged party to provide sufficient identification and authentication.

The identification protocol elements are passed between the parties with XID frames or X.25 registration packets. X.32 permits either method but requires that an identification exchange use only one method. Before explaining this protocol, several definitions are in order:

- Identity element (ID): A string of octets representing the identity of the challenged party

- Signature element (SIG): A string of octets representing the identity, such as a password, or a result of an encryption process

- Random number (RAND): An unpredictable range of octets used for each authentication

- Signed response element (SRES): Reply of the challenged party to the questioning party

- Diagnostic element (DIAG): Result of the identification process, transmitted by questioning party at the end of the identification process

X.32 stipulates two security options: (1) security grade 1 and (2) security grade 2. Security grade 1 involves two messages exchanged between the challenged and challenging party. Both options are in effect for the switched call. Any new identification process must be preceded by a disconnect of the link.

As shown in Figure 4.29a, the challenged party must first send its identity (ID) and, if required, some type of signature (SIG). In turn, the questioning party responds with the diagnostic (DIAG) message, which contains the result of the process. If it is not successful, security grade 1 permits up to three retries before a disconnect occurs.

The security grade 2 procedure involves a more enhanced authentication exchange (see Figure 4.29b). If the initial identification and signature of the challenged party is valid, the questioning party will return a message with a random number (RAND) which the challenge party must encrypt and return as its signed response (SRES). Also, the SIG could be used to convey the challenged party's public key or an indication that security grade 2 is to be used.

The questioning party then is required to decrypt the SRES. If the results of its decryption equals the value in the RAND, an appropriate diagnostic message (DIAG) is sent to the challenged party and the two parties have completed the authentication and identification process. If the identification process does not succeed, an error diagnostic message is returned and a connection is not made. X.32 permits only one attempt when the network is the questioning party.

The RSA public key algorithm. The security grade 2 option of X.32 is known as a *digital signature*. It is attractive because it makes use of the public cryptographic key concept. It is derived from the Rivest, Shamir, Adleman (RSA) public key algorithm.

With this approach, each user generates three keys: (1) a public modulo key (n), (2) a public exponential key (e), and (3) a private exponential key (d). The public keys can be revealed, but the private key must remain a secret and is known only to the user that generated it. Each encryption/decryption transformation takes the form:

$$X = Xk(\text{modulo } n)$$

where X is the value that is transformed (for example, RAND); X is the transformed value (for example, SRES), k is either e or d, n = (pq), which are large random prime numbers.

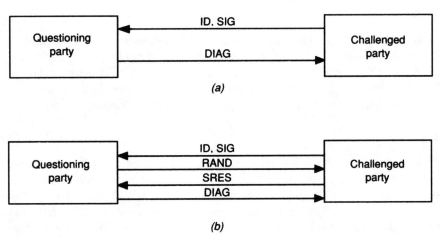

(a)

(b)

Figure 4.29 Security grades: *(a)* security grade 1; *(b)* security grade 2.

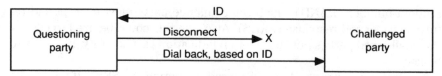

Figure 4.30   The X.32 secure dial-back feature.

Encryption and decryption both use the same modulo n key. Encryption uses either e or d as the exponential key k. Decryption uses either e or d as the exponential key k but must use the key opposite to that which was used in the encryption process.

The X.32 facilities

X.32 includes a commonly used feature for dial-in systems, the secure dial-back facility. This is an optional feature of X.32 which allows the sending DTE to identify itself and then be disconnected. In turn, the network uses the identity information provided in the dial-in procedure to dial-back the DTE. The DCE is required to identify itself and the DTE must once again identify itself. This is a very useful and relatively simple procedure and adds an additional level of security. It ensures the calling party is physically located at a registered public-switched network number. Figure 4.30 illustrates the X.32 dial-back feature.

One other facility is supported by X.32, the temporary location facility. It allows a registered DTE to designate to the network a dial-out number that is different from the registered number. Currently, the numbers are X.121-based and other plans are under study.

X.35

X.35 is a new recommendation added in 1993. It defines additional procedures for the operation of a gateway between private and public packet networks. Most of the X.25 procedures are used as defined in the X.25 Recommendation. However, because the two networks operate on peer-to-peer basis, a variety of minor modifications are made to the packet layer procedures.

X.38

The X.38 recommendation is a companion standard to X.5. It provides guidance for the interface of G3 facsimile equipment over a public data network with a facsimile packet assembly/disassembly (FPAD) facility. The recommendation defines the procedures for access to be provided from (a) the G3 facsimile equipment to an FPAD (b) or from the FPAD to the G3 facsimile equipment. In either case, the G3 equipment can be connected directly to the FPAD, or may be connected via GSTN to the FPAD.

The reader should note that the ITU-T publishes X.38 to operate with dual-toned multifrequency codes, which are defined in ITU-T Recommendation U.23 for signals defined in ITU-T Recommendations T.4 and T.30. A description of these recommendations are beyond this text. The interested reader should consult these documents for further information; otherwise, X.38 is difficult to comprehend.

The X.38 FPAD operates in a manner similar to a conventional X.3 PAD. The use of command signals and service signals are exchanged between the facsimile equipment and the FPAD to set up connections and select parameters for use during the session. In addition, state diagrams govern the behavior between the FPAD and the terminal equipment as well as timers and time-out values.

X.39

X.39 is similar to the packet-based PAD protocol X.29. Therefore, as the reader might expect, X.39 defines the procedures for the exchange of either control information or user data between an FPAD and a X.25 DTE or directly between FPADs.

For the packet mode DTE to FPAD communications, traffic exchange is in accordance with the X.25 Recommendation. For direct FPAD to FPAD communications, the Call Request packet must contain a destination telephone number of the G3 facsimile equipment. This number must be coded in accordance with ITU-T Recommendation X.121.

The FPAD messages are similar to the X.29 messages and contain operations such as SET, SET and READ, and READ. These messages are used to operate upon the X.5 parameters that reside in the FPAD. The signaling messages for the FPAD are coded in accordance with the ITU-T Recommendation T.30. These messages contain facsimile control fields such as multipage signals, end of message, end of transmission, partial page signal, failure to train, retrain positive, etc. It is obvious that the FPAD, as viewed by the ITU-T, is closely related to the ITU-T Recommendation T.30.

Summary

The ITU-T defines its network interfaces in Recommendations X.24 X.37. X.25 has become the dominant network interface throughout the world. The X.31 Recommendations are seeing increased use as telephone administrations and companies move to ISDN. X.32 is becoming the prevalent standard for dial-and-answer authentication protocols for packet- and circuit-switched networks. X.21 has seen extensive use in certain parts of the world. However, X.21 bis remains the prevalent interface for DTEs and DCEs. The X.28 and X.29 standards, although rather archaic, remain an integral part of a vendor's PAD products.

As suggested by their titles, these ITU-T Recommendations define a wide variety of operations for data communications networks. A number of these standards have remained relatively stable since they were written in the 1970s and 1980s, as have these standards that deal with telegraph transmissions and frequency-shift modulated systems. Some of the explanations in this chapter are terse, which reflects the brevity of the Recommendation. Note that these Recommendations are organized into "subcategories." As discussed in chapter 4, they are part of the public data networks recommendations. Figure 5.1a shows the organization of the Recommendations that deal with transmission, signalling, and switching. Figure 5.1b shows the organization of the Recommendations that deal with network aspects, maintenance, and administrative arrangements.

These standards include the following titles:

- X.40: Standardization of frequency-shift modulated transmission systems for the provision of telegraph and data channels by frequency division of a group (NLIF)

- X.50: Fundamental parameters of a multiplexing scheme for the international interface between synchronous data networks

- X.50 bis: Fundamental parameters of a 48 kbits user data signalling rate transmission scheme for the international interface between synchronous data networks

Public Data Networks

Transmission, signalling and switching

- X.50 Parameters for multiplexing between synchronous networks
- X.50 bis Parameters for 48-Kbit/s rates between synchronous nets
- X.51 Parameters for multiplexing with a 10-bit envelope structure
- X.51 bis Using 10-bit envelope structure on 48 Kbit/s interfaces
- X.52 Encoding anisochronous signals into a synchronous user bearer
- X.53 Numbering channels on multiplex links at 64 Kbit/s
- X.54 Allocation of channels on 64 Kbit/s multiplex links
- X.55 Using 6+2 envelope structure between synchronous networks
- X.56 Using 8+2 envelope structure between synchronous networks
- X.57 Transmitting a lower-speed channel on a 64 Kbit/s data stream
- X.58 Parameters for multiplexing with no envelope structure
- X.60 Common channel signalling for circuit switched applications
- X.61 SS No. 7-Data user part
- X.70 Start-stop control signalling between anisochronous networks
- X.71 Control signalling between synchronous networks
- X.75 Packet-switched signalling between public data networks
- X.80 Interworking of signalling for circuit-switched data services
- X.81 Interworking between ISDN CS and CSPDN
- X.82 Interworking between CSPDNs and PSPDNs based on T.70

Figure 5.1a The transmission, signaling, and switching aspects of the X Series.

- X.51: Fundamental parameters of a multiplexing scheme for the international interface between synchronous data networks using 10-bit envelope structure

- X.51 bis: Fundamental parameters of a 48 kbits user data signalling rate transmission scheme for the international interface between synchronous data networks using 10-bit envelope structure

- X.52: Method of encoding anisochronous signals into a synchronous user bearer

- X.53: Numbering of channels on international multiplex links at 64 kbits

- X.54: Allocation of channels on international multiplex links at 64 kbits

- X.55: Interface between synchronous data networks using a 6 + 2 envelope structure and single channel per carrier (SCPC) satellite channels

- X.56: Interface between synchronous data networks using an 8 + 2 envelope structure and single channel per carrier (SCPC) satellite channels

- X.57: Method of transmitting a single lower speed data channel on a 64 kbits data stream

- X.58: Fundamental parameters of a multiplexing scheme for the international interface between synchronous non-switched data networks using no envelope structure

- X.60: Common channel signalling for circuit switched data applications

- X.61: Signalling system No. 7-Data user part

- X.70: Terminal and transit control signalling system for start-stop services on international circuits between anisochronous data networks

- X.71: Decentralized terminal and transit control signalling system on international circuits between synchronous data networks

- X.75: Packet-switched signalling system between public networks providing data transmission services

- X.80: Interworking of interexchange signalling systems for circuit switched data services

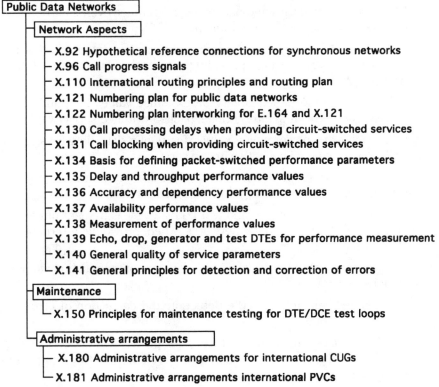

Public Data Networks

Network Aspects
- X.92 Hypothetical reference connections for synchronous networks
- X.96 Call progress signals
- X.110 International routing principles and routing plan
- X.121 Numbering plan for public data networks
- X.122 Numbering plan interworking for E.164 and X.121
- X.130 Call processing delays when providing circuit-switched services
- X.131 Call blocking when providing circuit-switched services
- X.134 Basis for defining packet-switched performance parameters
- X.135 Delay and throughput performance values
- X.136 Accuracy and dependency performance values
- X.137 Availability performance values
- X.138 Measurement of performance values
- X.139 Echo, drop, generator and test DTEs for performance measurement
- X.140 General quality of service parameters
- X.141 General principles for detection and correction of errors

Maintenance
- X.150 Principles for maintenance testing for DTE/DCE test loops

Administrative arrangements
- X.180 Administrative arrangements for international CUGs
- X.181 Administrative arrangements international PVCs

Figure 5.1b The network aspects, maintenance, and administrative arrangements of the X Series.

- X.81: Interworking between an ISDN circuit switched and a circuit switched public data network (CSPDN)

- X.82: Detailed arrangements for interworking between CSPDNs and PSPDNs based on Recommendation T.70

- X.92: Hypothetical reference connections for public synchronous data networks

- X.96: Call progress signals in public data networks

- X.110: International routing principles and routing plan for public data networks

- X.121: International numbering plan for public data networks

- X.122: Numbering plan for the E.164 and X.121 numbering plans

- X.130: Call processing delays in public data networks when providing international synchronous circuit-switched data services

- X.131: Call blocking in public data networks when providing international synchronous circuit-switched data services

- X.134: Portion boundaries and packet layer reference events for defining packet-switched performance parameters

- X.135: Speed of service (delay and throughput) performance values for public data networks when providing international packet-switched services

- X.136: Accuracy and dependability performance values for public data networks when providing international packet-switched services

- X.137: Availability performance values for public data networks when providing international packet-switched services

- X.138: Measurement of performance values for public data networks when providing international packet-switched services

- X.139: Echo, drop, generator, and test DTEs for measurement of performance values in public data networks when providing international packet-switched services

- X.140: General quality of service parameters for communication via public data networks

- X.141: General principles for the detection and correction of errors in public data networks

- X.150: Principles of maintenance testing for public data networks using data terminal equipment (DTE) and data circuit-terminating equipment (DCE) test loops

- X.180: Administrative arrangements for international closed user groups (CUGs)

- X.181: Administrative arrangements for the provision of international permanent virtual circuits (PVCs)

X.40

X.40 was published in Geneva in the 1972 recommendations and has not been altered since that date. Due to its antiquity, it is no longer in force (NLIF). It provides guidance for the use of frequency-shift modulated systems for telegraphs. One of the most interesting aspects of this specification is that it begins with "that some administrations are planning the introduction of public internetworks." This gives the reader an idea of how technologically dated some of the ITU-T Recommendations are. Of course, they need to be published because some countries still use them.

X.40 is a terse recommendation describing how to derive channels by frequency division within a group. It establishes specifications for modulation rates at 2400 and 9600 bauds. The 2400-baud channels use nominal mean frequencies at 110 kHz. The 9600-baud channels use a nominal mean frequency at 96 kHz for channel 1 and 72 kHz for channel 2.

X.40 provides other guidelines, such as defining the difference between the characteristics of the 2400-baud and the 9600-baud channels, the power requirements for the channels, the use of start and stop polarity frequencies, and so forth.

X.50

X.50 provides the specification for multiplexing on synchronous data networks. It was initially published in the 1972 recommendations, amended in 1976 and 1980, and has remained stable since that time. X.50 describes various alternatives for multiplexing envelopes. The term *envelope* describes the framing conventions for placing binary bits within one logical block or frame. The standard defines the coding conventions for 8-bit, multiple 8-bit, and 10-bit envelopes.

Control bits are also defined in X.50. Their specific use depends on the actual implementation. The reader can refer to X.30 in chapter 4 for a detailed explanation of the X.50 control bits and an example of a multiplexing scheme.

As an example of the use of X.50, the bit rate of 64 kbits is standardized for several interleaving options:

- 12.8-kbits channels repeat every fifth 8-bit envelope.
- 6.4-kbits channels repeat every tenth 8-bit envelope.
- 3.2-kbits channels repeat every twentieth 8-bit envelope.
- 800-bits channels repeat every eightieth 8-bit envelope.

Similar conventions are applied to the use of multiple 8-bit envelopes, as well as 10-bit envelopes.

X.50 bis

This ITU-T specification establishes the parameters for a 48-kbit transmission system used on synchronous networks. It employs the X.50 structures for (1) 8-bit envelopes and (2) 10-bit envelopes.

X.51

This recommendation is similar to X.50 with the exception that it provides the specifications for a multiplexing scheme in which both networks use a 10-bit envelope structure. X.51 specifies the use of the multiplex bit stream on a 64-kbit channel.

X.51 bis

This recommendation was written in 1980 and has not gone through any changes since that date. It is a very short specification defining the requirements for a 48-kbit signalling rate using a 10-bit envelope structure.

X.52

X.52 defines the procedures for encoding anisochronous signals into a synchronous user channel. It defines the encoding method for the characters generated by the DTEs in X.1 classes of service 1 and 2. It defines the start and stop polarities for the signalling, and it states that there is no need to have any relation between the characters in the envelopes.

X.52 requires that a multiplexing encoder be implemented in a manner that the time delay between the reception of a character and the start of sending the character on a channel is less than 1 bit of the data rate of the actual synchronous user channel.

X.53

X.53 is a recommendation of approximately one-half page in length that describes how to number the channels on multiplex links at 64 kbits. It uses the concepts defined in X.50 and X.51.

X.54

X.54 defines the configuration and phase numbers for multiplex channels operating at 64 kbits. This specification consists of one table that describes the configuration number and one of five phase numbers, and the speeds between the two.

X.55

X.55 describes the interface between two synchronous networks using a 6+2 envelope structure and single-carrier-per-channel satellite circuits. It was added to the Red Book in 1984. Even though ITU-T recognizes that a basic carrier rate channel is 64 kbits, this specification was based on the fact that the 64-kbit technology was not available on some satellite systems. Consequently, for an interim period this specification developed schemes for the use of 48-, 50-, or 56-kbit channels. It describes the adaptation procedures between a 56-kbit and a 64-kbit channel, using X.50 as the reference.

X.56

This recommendation is similar to X.55 except that it provides the definitions for using an 8+2 envelope structure on a single-carrier-per-channel satellite circuit.

X.57

X.57 describes the methods for transmitting a lower-speed channel on a 64-kbit data stream. It makes reference to X.30 for the adaptation of synchronous user data rates on a 64-kbit channel. It stipulates that the 6+2 envelopes that relate to the low-speed channels shall be repeated as many times as necessary to reach the 64-kbit speed.

X.58

This recommendation is similar to X.50 except that it defines a scheme for multiplexing with nonenvelope structured data.

X.60

X.60 is one-half page in length. It states that the common channel signalling for a circuit-switched data application shall be in accordance with signalling system no. 7.

X.61

This recommendation is used in conjunction with the signalling system no. 7 (SS7) recommendations. It defines the call control facility, registration, and cancellation signals needed for international common-channel signalling (assuming, of course, that SS7 is employed). It refers to X.300 for the use of user and network facilities.

X.61 describes the use of signalling messages such as call and setup messages, addressing messages, call accepted and rejected messages, clearing messages, and so forth. The recommendation also defines in considerable detail the signal information that is transferred in the messages. As examples, the components inside the messages can include (but are not restricted to) destination codes, originating codes, heading information such as field indicators, data country code (DCC) and data network identification code (DNIC) addresses, alternate routing indicators, and typical telephone messages such as busy signals, access barred signals, change number signals, out-of-order signals, etc. The user should refer to X.61 for more information on these message components. They are quite numerous.

X.61 also defines the formats and codes for the data signalling messages. These are referred to as signal units (SUs). The ITU-T Recommendation Q.703 (message transfer part) describes the formats for these SUs.

Section 4 of X.61 defines the call control procedures to be used on circuit-switched data transmission systems. These definitions are dependent on the characteristics of the DTE-DCE interfaces that are in operation on the system. Two phases are defined: (1) call setup and (2) call clear-down. Section 4 describes only basic calls and does not describe any additional user facilities; the facilities are found in X.300.

Generally, the call setup proceeds as follows: The originating exchange must receive the complete selection information from the calling user before it begins its operations. On receiving this information, if it determines that the call is to be routed to another exchange, it must first seize a free data circuit and then send an address message on the link. This message must contain all the information that is required to set up the call to the called user.

The address message is received by a transit exchange, which analyzes the destination address to determine the routing of the call. It, in turn, seizes a free interchange circuit and sends the message to the next exchange until the message reaches the end destination.

On receiving the address message, the destination exchange analyzes the address to determine which user should receive the call. As with other telephone-type systems, it checks the user's line condition and other operations to determine if the call is going to be allowed. If allowed, the exchange will then call the receiving user in accordance with the specific interface protocol existing at the DTE-DCE. Ordinarily, the call user will respond with the call accepted or a similar signal.

The remote destination exchange ordinarily will send a call accepted message to the preceding exchange; it will then be relayed back to the originating exchange. When this call accepted message is received, the originating exchange then prepares to connect through the data path. It performs a connection and starts charging, if appropriate.

The call clear-down follows a similar procedure except that a valid clearing signal is required from the clearing user. Either the originating or destination user can clear. Whichever station clears, the originating or destination exchange will release the connection and send a clear message back to the adjoining exchange. This exchange message is routed to the other communicating exchange. The clearing may also be initiated by an exchange because of some problems or network conditions.

For more detailed information on SS7, the reader should read the ITU-T Q Series.

X.70

This recommendation defines the operations for start-stop terminals operating on international networks. ITU-T considers that two classes of service (class 1 and class 2) are applicable to anisochronous types of networks that require a data rate of 300 and 200 bits, respectively.

This standard contains a number of tables and diagrams that define the control and call progress signals for these types of terminals.

X.71

This recommendation is similar to X.70 except, as the name implies, it deals with synchronous systems. X.71 is principally a document consisting of tables and state diagrams that are used to specify the call setup between terminals on synchronous networks. For purposes of documentation, the letter x denotes the side that originates the call, and the letter y denotes the side that receives the call. Using these letters, X.71 describes the forward path of x toward y and the backward path of y toward x and then describes the signals that move on these paths.

X.75

Introduction

X.25 is designed for users to communicate with each other through one network. However, two users operating on two separate X.25 networks may need to establish communications to share resources and exchange data. X.75 is designed to meet this need. It is also used to connect packet exchanges within a network. Moreover, X.75 is used as an internetworking protocol between ISDNs and packet networks (see X.31). The recommendation has been in development for almost 10 years; it was published as a provisional recommendation in 1978, amended in 1980, and amended again in 1984 and 1988.

X.75 is quite similar to X.25. It has many of its features, such as permanent virtual circuits, virtual call circuits, logical channel groups, logical channels, and several of the control packets. The architecture is divided into physical, link, and packet levels, which is also quite similar to X.25.

X.75 defines how two terminals are connected logically by an international link while each terminal is operating within its own packet-mode data network. X.75 uses a slightly different term for the network interface. In the description of X.25, the term DCE is used to describe an X.25 packet exchange. The X.75 terminology defines this device as a signalling terminal (STE), even though it may be the same as the X.25 device.

X.75 requires the physical signalling to be performed at 64 kbits and an optional rate of 2.048 Mbits. Of course, many vendors use other link speeds (56 kbits, 1.544 Mbits, etc.). With a few minor changes, the second level of X.75 uses the high-level data link control (HDLC) subset link access procedures balanced (LAPB). X.75 does not support LAP.

X.75 does not use the variety of packet types that are found in X.25, primarily because the STE-to-STE communications has no relationship to the "other side of the cloud." In X.75, there is no other side, because the communication is only between two STEs and not two sets of DCEs and DTEs.

Figure 5.2 shows the operations of X.75 between networks A and B and networks B and C. In addition, the X.25 operations are shown at the DTE-

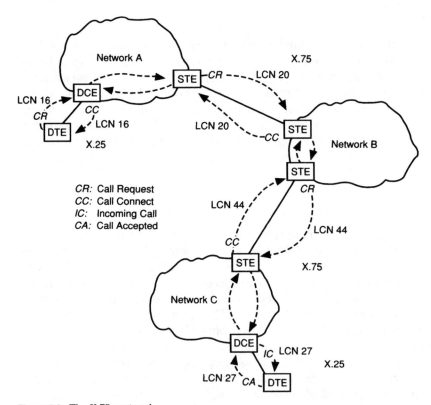

Figure 5.2 The X.75 protocol.

DCE interfaces for networks A and C. Notice that X.75 needs only the Call Request and Call Confirmation packets; it does not use the Incoming Call or Call Accepted packets.

X.75 utilities

The X.75 packet carries an additional field called network utilities. Its purpose is to provide for network administrative functions and signalling. In many situations, a request in the X.25 user packet facility field invokes the use of an X.75 network utility. The X.25 facilities that do not require any STE action are relayed transparently through the STE. Other user facilities that require STE action are mapped into the X.75 utilities field. These utilities are listed in Table 5.1, and each is described as follows:

- Call identifier: An identifying name established by the originating network for each virtual circuit. When used in conjunction with the DTE address, it uniquely identifies the call (each virtual circuit).

- Called line address modified notification: Because of Hunt groups and call redirections, this utility identifies the specific reason for the called address to be different from the address in the Call Request packet.

- Clearing network identification code: Provides additional information on the origin of the Clear Request packet.

- Closed user group indication: Used to enable the establishment of calls between DTEs that are members of an international closed user group. X.75 also supports a closed user group with outgoing access indication.

- Fast select indication: Indicates that a fast select is requested for the call.

- Packet size indication: Identifies the negotiated packet size between the STEs.

- Reverse charging indication: Allows reverse charging of calls to be established across the networks.

- RPOA selection: Used to select a transit network or networks to handle the call. Recognized private operating agencies (RPOAs) are discussed in the X.2 Recommendation.

- Tariffs: Used for billing and accounting purposes. The content of this field is determined by the originating and destination network.

- Throughput class indication: Indicates the throughput classes applying to the call.

- Traffic class indication: Identifies service information, such as terminal, facsimile, etc. This utility has not yet been fully defined.

- Transit delay indication: Used by the network to identify the accumulated transit delay on the virtual circuit. This parameter is defined in X.135 as t3.

- Transit delay selection: Used by the DTE to request a transit delay value. It may be used with the transit delay indication facility to make routing decisions.

- Transit network identification: Contains the first four digits of the international data number (DNIC or INIC, see X.121) of the transit network controlling a portion of the virtual circuit. If more than one transit network is involved in the call, the DNIC or INIC of each network is stored in this field. Thus, this field provides information on the networks that handle the call. This identification is also present in the Call Connected packet.

- Utility marker: Used to separate X.75 utilities from non-X.75 utilities. Its use is subject to bilateral agreements between networks.

- Window size indication: Identifies the negotiated window size between the STEs

X.75 and multilink procedures (MLP)

The X.75 link level frequently uses the MLP. This procedure provides for the use of multiple links between STEs. MLP establishes the rules for frame transmission and frame resequencing for delivery to and from the multiple links. Multilink operations allow the use of parallel communications channels between STEs in such a manner that they appear as one channel. The multilink operation provides for more reliability and throughput than what can be achieved on a single channel.

MLPs exist at the upper part of the data link level. The X.25 network layer perceives it as connected to a single link and the LAPB single links operate as if they are connected directly to the network layer. MLP is responsible for flow control between layers 2 and 3, as well as resequencing the data units for delivery to the network layer.

TABLE 5.1   The X.75 Utilities

Call identifier
Called line address modified
Clearing network identification code
Closed user group indication
Fast select indication
Packet size indication
Reverse charging indication
RPOA selection
Tariffs
Traffic class indication
Transit delay indication
Transit delay selection
Transit network identification
Throughput class indication
Utility marker
Window size indication

X.80

This recommendation defines the interworking of systems that use the X.60, X.70, and X.71 circuit-switching standards. X.80 consists of several flow charts and tables that describe the required signals to set up and take down the call. The flow charts are based on ITU-T's specifications and descriptions language (SDL), which is explained in Recommendations Z.101 and Z.103.

X.81

This recommendation describes the arrangements for internetworking between CSPDNs and ISDNs. It is a complement to and based on the X.300 Recommendations. The specific arrangements for call control between the CSPDN and the ISDN are defined in Recommendation X.301. The facilities are defined in X.302.

X.81 stipulates that the interworking functions can be provided by either the ISDN or the CSPDN. The links may be provided as multiplexed lines or as individual links. Multiplexing is actually stipulated based on the location of the interworking function (IWF).

Because a circuit-switched data network and an ISDN may have different bit signalling rates, X.81 provides a bit rate adoption mechanism by citing the use of Recommendation X.30 to adapt the 40-bit ISDN frames to either 6+2 (X.50) or 8+2 (X.51) envelopes.

X.81 defines the procedures for the support of the OSI network layer service (OSI-NLS). This is done also in accordance with X.300. In addition, it may be necessary to provide signalling conversion (also known as protocol mapping) from an SS7 ISDN user part to the X.71 signalling scheme or the SS7 based on X.60, X.61, or Q.761 to Q.766.

X.82

As the reader might expect, X.82 adheres to X.300. ITU-T publishes this recommendation principally to be applicable to interworking that involves telematic services and not to the interworking involving communications capabilities generally described in X.300. The specification outlines several scenarios for the mapping of the circuit-switched X.71-oriented signals through the interworking facility based on T.70 to the packet-switched data network using X.75.

X.92

This recommendation establishes the specific connections available for DTEs into data networks. The following links are permitted within the standard:

- Link A: Data link between two adjacent data switching exchanges in a national network

- Link A1: Data link between two adjacent gateway data switching exchanges (DSE) in an international connection

- Link B1: Data link between a local DSE and a gateway DSE

- Link G1: Data link between a source gateway DSE and a destination gateway DSE in an international connection

- Link C: Data link between source DTE and destination DTE

- Link D: Data link between source DTE and the source local DSE or the data link between destination DTE and destination local DSE

- Link E: Data link between communicating processes

X.25 and X.75 use the X.92 Recommendation for their links and reference connections.

X.96

This standard establishes the signals that may be used to inform DTEs (such as calling DTEs) of the progress of a connection call or connection request through a public network. X.96 defines the signals to be returned to the caller to indicate connections that were not made (and why) and to indicate circumstances regarding the progress of a call through a network. It can be very valuable for a calling DTE to know if (1) there is a problem detected at the DTE-DCE interface, (2) if a virtual call has been reset or cleared, or (3) if a permanent virtual circuit has been reset. The entries in Table 5.2 pertaining to X.96 are relevant to a packet-switched network. The following symbols are used in the table:

M = Mandatory
(M) = Mandatory when relevant facility is used
FS = For Further Study
- = Not applicable

TABLE 5.2   X.96 Call Progress Signals

Call progress signal	Definition	Category	VC	PVC
Registration/ cancellation confirmed	The facility registration or cancellation requested by the calling DTE has been confirmed by the network.	B	(M)	(M)
Local procedure error	A procedure error caused by the DTE is detected by the DCE at the local DTE-DCE interface. Possible reasons are indicated in relevant Series X interface recommendations.	D1	M	M
Network congestion	A congestion condition exists in the network	C2	M	M

TABLE 5.2   X.96 Call Progress Signals (Continued)

Call progress signal	Definition	Category	VC	PVC
Network out of order	Temporary inability to handle data traffic.	C2	-	M
Invalid facility request	A facility requested by the calling DTE or the called DTE is detected as invalid by the DCE at the local DTE-DCE interface.	D1	M	-
RPOA out of order	The RPOA nominated by the calling DTE is unable to forward the call.	D2	(M)	-
Not obtainable	The called DTE address is out of the numbering plan or not assigned to any DTE.	D1	M	-
Access barred	The calling DTE is not permitted the connection to the called DTE.	D1	M	-
Reverse charging acceptance not subscribed	The called DTE has not subscribed to the reverse charging acceptance facility.	D1	(M)	-
Fast select acceptance not subscribed	The called DTE has not subscribed to the fast select acceptance facility.	D1	(M)	-
Incompatible destination	The remote DTE-DCE interface or the transit network does not support a function or facility requested.	D1	M	M
Ship absent	The called ship is absent.	D1	M	-
Out of order	The remote number is out of order. Possible reasons include: DTE is uncontrolled not ready DCE power off Network fault in the local loop In packet switches only (X.25 level 1 not functioning, X.25 level 2 not in operation).	D1 or D2	M*	M*
Network fault in the local loop	The local loop associated with the called DCE is faulty.	D2	*	*
DCE power off	Called DCE has no main power or is switched off.	D1	*	*
Uncontrolled not ready	Called DTE is uncontrolled not ready.	D1	*	*
Controlled not ready	Called DTE is signalling controlled not ready.	D1	FS	FS
Number busy	The called DTE is detected by the DCE as engaged on other call(s) and therefore as not being able to accept the incoming call.	C1	M	-
Remote procedure error	A procedure error caused by the DTE or an invalid facility request by the remote DTE is detected by the DCE at the remote DTE-DCE interface. Possible reasons are indicated in relevant Series X interface recommendations.	D1	M	M

TABLE 5.2 X.96 Call Progress Signals (Continued)

Call progress signal	Definition	Category	VC	PVC
Long-term network congestion	A major shortage of network resource exists.	D2	-	-
Network operational	Network is ready to resume normal operation after a temporary failure or congestion.	C1	-	M
Remote DTE operational	Remote DTE-DCE interface is ready to resume normal operation after a temporary failure or out of order condition (e.g., restart at the DTE-DCE interface). Loss of data may have occurred.	C1	-	M
DTE originated	The remote DTE has initiated a clear, reset, or restart procedure.†	B or D1	M	M
PAD clearing	The call has been cleared by the local PAD as an answer to an initiation from the remote DTE.	B	M (X.28 only)	

*Although the basic out-of-order call progress signal is transmitted for these conditions, the diagnostic field in the clearing or resetting packet may give more precision.

†Possible reasons for this include reverse charging not accepted.

X.110

Recommendation X.110 establishes several routing principles for circuit-switched and packet-switched calls on data networks. Its intent is to foster international procedures for internetworking between national networks and international data switching exchanges (IDSEs). Its value is in its definitions and general examples. It does not contain enough detailed information for any meaningful detailed design decisions.

X.121

This recommendation has received considerable attention throughout the world because its intent is to provide a universal addressing scheme, allowing users to communicate with each other through multiple networks. X.121 establishes a standard numbering scheme for all countries' networks and individual users within those networks.

A DTE within a public data network is addressed by an international data network address. The international data network address consists of a DNIC and a network terminal number (NTN). Another option is to provide the international data number as the DCC and a national number (NN).

The four codes consist of the following identifiers. The DNIC consists of four digits; the first three digits identify the country and can be regarded as DCC. The fourth digit identifies a specific data network within a country. The network terminal number can consist of 10 digits or, if used in place of the NTN, an NN is used and then 11 digits are allowed.

Important parts of X.121 are the descriptions of prefixes and escape codes. A prefix (P) is a value of one or more digits that allows the use of different types of address formats. It precedes the X.121 address, although it is not considered part of the number and is not transmitted over an internetworking boundary. An escape code is a one-digit value that indicates the address in a number from a different address plan. It is carried across an internetworking boundary. Most escape code implementations use the values of 0, 8, or 9.

The X.121 international numbering scheme is summarized as follows (the numbers in parentheses indicate the number of digits in each field):

```
P  +  DNIC  +  NTN
(1)     (4)      (10)
```

or

```
P  +  DCC  +  NN
(1)     (3)     (11)
```

Although the X.121 scheme permits the assignment of 600 country codes and 6000 DNICs, it has proved inadequate to support a country in which more than 10 networks need to be identified. To overcome this problem, some countries are assigned more than one DCC. For example, the United States is assigned the DCC values of 310 through 316.

X.122

The X.122 Recommendation was added in the Blue Book and is a very useful guide on numbering plans for internetworking between packet networks, ISDNs, and telephone networks. It is based on many of the concepts in X.121 and provides for two escape codes for internetworking systems that use the X.121 and E.164 numbering plans. Escape code 0 is used to escape from the X.121 plan to the E.164 plan, with the indication that a digital interface exists between the packet network and the destination ISDN or telephone network. Escape code 9 is used to escape from and to the same plans, with the indication that an analogue interface is used.

In case the escape codes 0 and 9 are used by a network for other operations, X.122 specifies the use of an optional network-specific digit(s) (ONSD). This value must provide the same functionality as the escape codes, and the translation of this address to the escape code is the responsibility of the originating or transit network.

Figure 5.3 provides examples of X.122 rules of address mapping between packet networks, ISDNs, and telephone networks. The terms CD and CG mean called address and calling address, respectively.

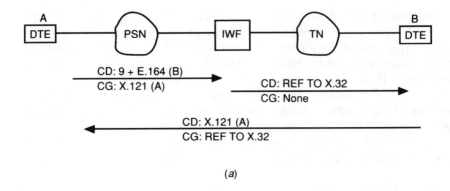

(a)

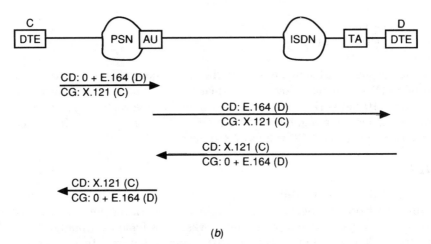

(b)

Figure 5.3 X.122 address mapping examples. (a) packet-switched network (PSN) to and from the telephone network (TN); (b) packet-switched network (PSN) to and from ISDN.

X.130

X.130 specifies the definitions and parameters for call processing delays in international synchronous circuit-switched networks. In defining the total call connection delay, X.130 defines the parts of the connection for measuring the delays as shown in Figure 5.4.

The total call connection delay (TCCD) is measured by the sum of six elements:

- The delay between the call request signal transmitted by the calling DTE and its receipt of the proceed-to-select signal from the local DSE

- The time between the receipt of the proceed-to-select signal and the transmission of the end-of-selection signal by the calling DTE
- The time between the transmission of the end-of-selection signal by the calling DTE and the receipt of the incoming call signal at the remote DTE
- At the remote called DTE, the delay between the receipt of the incoming call signal and its transmission of the call accepted signal
- The delay between the sending of a call accepted signal at the remote DTE and the receipt of the ready-for-data signal at the local DTE
- At the remote DTE, the delay between the transmission of the call accepted signal and the receipt of the ready for data signal from the DSE

After establishing these definitions, X.130 provides statistics on acceptable delay values. It also provides values as well as definitions for the clearing operations.

X.131

This recommendation provides definitions and guidance for using the X.21 and X.21 bis interfaces on an international synchronous circuit-switched network. Its specific function is to establish the values for call blocking probability. The blocking probabilities are specified for normal or busy hour loads.

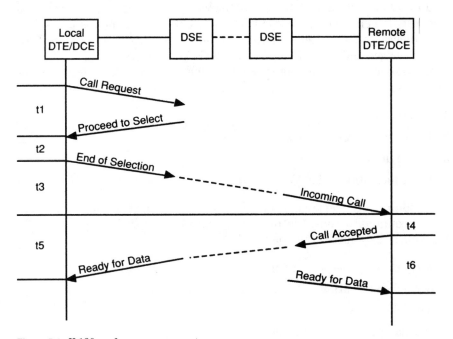

Figure 5.4 X.130 performance parameters.

X.131 uses some of the documentation in X.92 for defining two types of connections:

■ Type 1: Terrestrial connections without satellite circuits

■ Type 2: Interconnections with satellite circuits in either the national or international portion of the connection

X.131 states that the overall probability of nonservice due to congestion should not exceed 13 percent for connection type 1 and 15 percent for connection type 2.

X.134, X.135, X.136, X.137, X.138, and X.139

These ITU-T Recommendations provide guidance on defining and using performance parameters and values on a packet-switched network. Many networks use these specifications to support their X.25- and X.75-based products. Figure 5.5 shows the structure and relationships of these four recommendations. The documents are named as follows.

X.134 sets the framework for the other specifications. It contains information on the boundaries for national and international virtual circuit connections and establishes the monitor points for conducting performance analysis. It also provides a detailed description of the performance-significant events resulting from the use of the various X.25 packets, such as Call Request, Receive Not Ready (RNR), etc. X.134 also contains a detailed state diagram that explains the rules for flow control across the X.25 DTE-DCE interface and the X.75 STE-X and STE-Y interface.

X.135 specifies the parameters and tools for measuring delay and throughput across multiple X.25 and X.75 networks. It defines four parameters:

1. Call set up delay

2. Data packet transfer delay

3. Throughput capacity

4. Clear indication delay

X.136 provides the specifications to define and measure network blocking. It contains permissible probability thresholds for the following X.25 blocking situations:

1. Virtual call request rejection

2. Virtual call clearing

3. Virtual call reset

4. Permanent virtual call reset

5. DTE restart

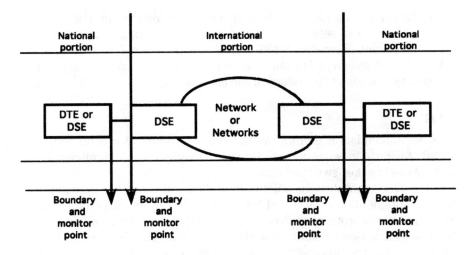

X.135, X.136, x.137, X.138, and X.139:
speed, accuracy, dependability, and
service availability

Figure 5.5 X.134, X.135, X.136, X.137, X.138, and X.139 functional view.

As an example of how X.136 is applied, the probability that a DTE receives a Clear Indication packet because the network path is not available is not to exceed 9×10^{-3}.

X.137 uses eight performance parameters to compute the availability of an X.25 or X.75 virtual connection. Several formulas are provided to calculate mean time between service outages (MTBSO), mean time to service restoration (MTTSR), failure rate, and unavailability. The performance parameters are:

1. Call set up failure probability
2. Call set up error probability
3. Throughput capacity
4. Residual error rate
5. Reset probability
6. Reset stimulus probability
7. Premature disconnect probability
8. Premature disconnect stimulus probability

X.138

I have found the X.138 and X.139 to be quite helpful to me in some of my ongoing work with clients. When used in conjunction with the other X.130

Series (X.130 through X.137), these specifications provide practical guidelines on how to develop a methodology for measuring performance in packet-switched networks. X.138 established the overall architecture for this methodology, several calculations of packet performance statistics, and the factors that affect the values of the calculations.

X.139

This recommendation defines formula and procedures for estimating the speed of service in relation to delays and throughput through international and national packet-switched data networks.

X.139 also establishes the points where the monitoring occurs to perform the measurements for delay and throughput. Therefore, at certain boundaries, monitors (protocol analyzers) are placed to determine call set up delay between various points on a virtual connection. These performance factors include the processing of Call Request packets, Incoming Call packets, Call Accepted packets, and Call Connected packets. You may know now that X.139 assumes that the connection being measured is based on the X.25 Recommendation. Therefore, if you are faced with establishing and measuring X.25 performances through one network or multiple networks, X.139 will be a very valuable reference guide. The X.139 specification (unlike some of the ITU-T specifications) is quite pragmatic, and contains algorithms showing how to measure the speed of service through packet data networks.

Where to find the information

The reader is encouraged to study all the X.130 Recommendations as a whole, because they are interdependent. Table 5.3 provides a guide to the specific discussions and performance parameters and values.

TABLE 5.3   Guide to X.134 through X.139

Call set-up delay	X.135, section 2, X.138 Annex A
Data packet transfer delay	X.135, section 3, X.138 Annex A
Throughput	X.135, section 4, X.138 Annex A
Clear indication delay	X.135, section 5, X.138 Annex A
Clear confirmation delay	X.135, section 5
Call set-up error probability	X.136, section 2, X.138 Annex A
Call set-up failure probability	X.136, section 2, X.138 Annex A
Residual error rate	X.136, section 3, X.138 Annex A
Reset stimulus probability	X.136, section 3, X.138 Annex A
Reset probability	X.136, section 3, X.138 Annex A
Premature disconnect stimulus probability	X.136, section 3, X.138 Annex A
Premature disconnect probability	X.136, section 3, X.138 Annex A
Call clear failure probability	X.135, section 4, X.138 Annex A
Availability	X.135, section 4, X.138 Annex A
Mean time between service outages	X.137, X.138 Annex A

X.140

X.140 defines the quality-of-service (QOS) parameters to measure end-to-end QOS when using data networks. These QOS parameters focus on (1) performance effects that are measurable at the network interfaces and not within the network and (2) the definitions that are based on protocol-independent events (that is, they focus on, say, an access request rather than the issuance of an X.21 or X.25 Call Request packet).

X.140 is a very useful reference guide in that it shows the relationship of the general parameters to the following specific parameters:

- Circuit-Switching Parameters: X.130 and X.131
- Packet-Switching Parameters: X.135 and X.136
- OSI Network Layer Service Parameters: X.213

The following general QOS parameters are defined in X.140:

- Access Delay: The time lapsed between an access request and the successful access
- Disengagement Delay: The elapsed time between the beginning of a disengagement attempt and the successful disengagement
- User Information Transfer Delay: The elapsed time between the start of transfer and successful transfer of a user protocol data unit
- Access Denial Probability: The ratio of total access attempts resulting in access denial to the total access attempts during a specified sample
- Incorrect Access Probability: The ratio of total access attempts resulting in incorrect access to total access attempts during a specified sample
- Disengagement Denial Probability: The ratio of total disengagement attempts that result in disengagement denial to total disengagement attempts in a specified sample
- User Information Loss Probability: The ratio of total lost user information units to the total transmitted user information in a specified sample
- User Information Transfer Denial Probability: The ratio of total transfer denials to total transfer samples during a specified observation period
- User Information Misdelivery Probability: Ratio of total misdelivered user information units to total user information units transferred between a specified source and destination user in a specified sample
- Extra User Information Delivery Probability: The ratio of total (unrequested) extra information units to total information units received by a destination user in a specified sample
- Service Outage Duration: The duration of any continuous period of time for which satisfactory or tolerable service is not available

- Service Availability: The ratio of the aggregate time during which satisfactory or tolerable service is or could be provided to the total observation period

- User Information Error Probability: The ratio of total incorrect user information units to the total successfully transferred user information units plus incorrect user information units in a specified sample

- User Information Transfer Rate: The total number of successfully transferred user information units in an individual transfer sample divided by the input-output time for that sample

X.141

X.141 establishes the principles for the detection and correction of errors that occur in data networks. It is based on using two types of error control operations:

- Forward error correction (FEC): Corrects the error at the receiver instead of retransmitting the data

- Automatic request for repeat: Detects an error and requests a retransmission from the transmitter

Both these techniques have been discussed in previous sections of this book. For forward error detection the reader should refer to Recommendations V.32 and V.33. For automatic request for repeat the reader may refer to the material in this book dealing with HDLC, LAPB, and link access procedure for the D channel (LAPD). Also, X.141 contains some useful information on the use of reject (REJ) and selective reject (SREJ) in satellite links.

X.150

X.150 establishes some general principles for loopback testing operations in DTEs and DCEs. The loopbacks defined in this recommendation are shown in Figure 5.6.

X.180

X.180 is a very terse document that provides guidance for the administration of international closed user groups (ICUGs). It gives the procedures for applying for a closed user group number with the coordinating administration.

X.181

X.181 is another very brief standard that defines the administrative procedures for providing international PVCs. As with X.180, it defines the inter-

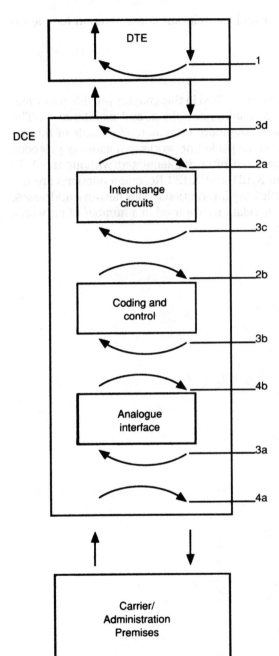

Figure 5.6 Loop tests with X.150.

action between the subscriber and the relevant administration for the establishment of the PVC.

Summary

The X Series Recommendations described in this chapter provide many features pertaining to network signalling, switching, and maintenance. The X.75 Recommendation is one of the more prominent standards in this series. It is used extensively in public packet networks as a gateway protocol. The X.75 multilink procedures are often implemented without the X.75 packet level procedures. The X.121 and X.122 Recommendations are the prevalent standards for establishing international data network addresses. The X.134 and X.137 Recommendations are used in a number of networks for defining performance.

6

The X.200 Series

The Open Systems Interconnection (OSI) Model was developed by several standards organizations (including the ITU-T) and is now a widely used model for the design and implementation of computer networks. It was introduced in chapter 1 under "The X Series and the OSI Model" and in several sections in chapter 2. The reader should review this material before reading this chapter.

OSI is intended to diminish the effects of the vendor-specific approach to the design of communications systems which has resulted in each vendor system operating with unique protocols. This approach has "closed" end users to options of interconnecting and interfacing with other systems and has necessitated the purchase of expensive and complex protocol converters in order to "translate" the different protocols.

The OSI X.200 Series has been in development for about 10 years. The 1984 Red Book provided considerable guidance on how to implement the OSI Model under the X Series framework. However, several protocols and service definitions were missing from the Red Book because they were not complete at the time of publication. Most of the pieces of the puzzle are now complete with the 1988 Blue Book, and subsequent Recommendation.

The ITU-T publishes its OSI Model specifications in the X.200 X.290 Recommendations. The X.200 documents contain slightly over 1100 pages. It is impossible to examine these recommendations in one chapter in any detailed fashion. Therefore, the goal of this chapter is to provide the reader with a general foundation of the ITU-T OSI Model. Another book is available for the reader who wishes more information on the OSI Model, *OSI: A Model for Computer Communications Standards,* by Uyless Black, Prentice-Hall, 1991.

Format for this chapter

Figures 6.1*a* and 6.1*b* show the general organization of the X.200 Series, and their titles are presented in the list following. The reader should be aware that my approach in describing the X Series Recommendations in this chapter differs from the previous chapters. Because these Recommendations are organized on the seven-layer OSI Model concept, they are ex-

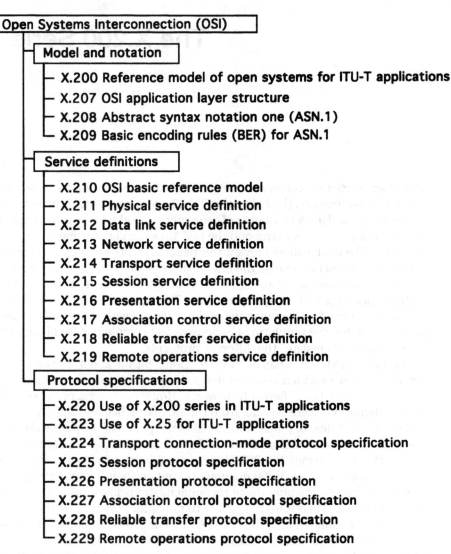

Figure 6.1*a* OSI model and notation, service definitions, and protocol specifications of the X Series.

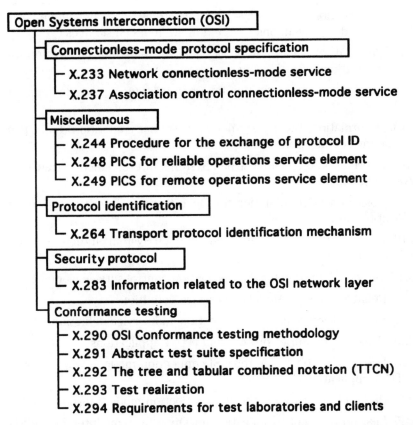

Figure 6.1b Other parts of the X.200 Recommendations (See chapter 13 for X.290 – X.294).

amined in that context, rather than on an individual one-to-one description. The X.200 Recommendations that do not pertain to any one layer are discussed in another section. The table of contents at the beginning of this book reflects this approach. Also, Recommendations X.290 through X.294 are covered in chapter 13.

- X.200: Reference Model of Open Systems Interconnection for ITU-T Applications

- X.207: Information technology—Open Systems Interconnection—Application layer structure

- X.208: Specification of Abstract Syntax Notation One (ASN.1)

- X.209: Specification of basic encoding rules for Abstract Syntax Notation One (ASN.1)

- X.210: Information technology—Open Systems Interconnection-basic reference model-conventions for the definition of OSI services
- X.211: Physical service definition of Open Systems Interconnection for ITU-T applications
- X.212: Data link service definition for Open Systems Interconnection for ITU-T
- X.213: Information technology-network service definition for Open Systems Interconnection
- X.214: Information technology–Open Systems Interconnection-transport service definition
- X.215: Session service definition for Open Systems Interconnection for ITU-T applications
- X.216: Presentation service definition for Open Systems Interconnection for ITU-T applications
- X.217: Service definition for the association control service element
- X.218: Reliable transfer: Model and service definition
- X.219: Remote operations: Model, notation and service definition
- X.220: Use of X.200-Series protocols in ITU-T applications
- X.223: Use of X.25 to provide the OSI connection-mode network service for ITU-T applications
- X.224: Protocol for providing the OSI connection-mode transport service
- X.225: Session protocol specification for Open Systems Interconnection for CCITT applications
- X.226: Presentation protocol specification for Open Systems Interconnection for CCITT applications
- X.227: Connection-oriented protocol specification for the association control service element
- X.228: Reliable transfer: protocol specification
- X.229: Remote operations: protocol specification
- X.233: Information technology—Protocol for providing the connectionless-mode network service: protocol specification
- X.237: Connectionless protocol specification for the association control service element
- X.244: Procedure for the exchange of protocol identification during virtual call establishment on packet switched public data networks
- X.248: Reliable transfer service element—Protocol implementation conformance statement (PICS) proforma

- X.249: Remote operations service element—Protocol implementation conformance statement (PICS) proforma
- X.264: Transport protocol identification mechanism
- X.283: Elements of management information related to the OSI network layer

Key Terms and Concepts

The ITU-T uses two key terms in describing the X.200 Series Recommendations:

Service definitions: Defines the services and operations that take place between the layers of the Model within the same machine. The service definitions are implemented with primitives.

Protocol specifications: Actions taken within or between peer layers of the Model across different machines (or two peer layers within the same machine). These actions taken are based on the service definitions.

Figure 6.2 shows the relationships of the service definitions and protocol specifications. The transport and network layers are chosen for this example, although all seven layers behave in the same manner.

Through the use of four types of service definitions, called *primitives* (request, indication, response, and confirm), the adjacent layers in a machine coordinate and manage the communications process. Some sessions do not use all four types; this example shows the request (req) and indication (ind) primitives. The layers can invoke different kinds of these primi-

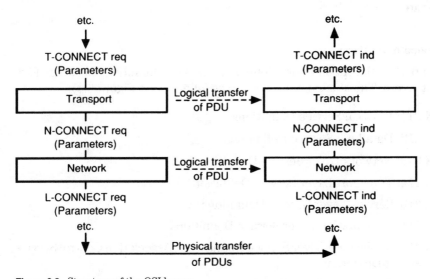

Figure 6.2 Structure of the OSI layers.

tive types. In this example, CONNECT primitives are illustrated because a connection-oriented service is being requested. Other types could be used, such as DATA, DISCONNECT, and ABORT primitives. The primitive types perform the following functions:

Request: Primitive by service user (the adjacent upper layer) to invoke a function

Indication: Primitive by service provider (the adjacent lower layer) to indicate a function has been invoked (either successfully or unsuccessfully)

Response: Primitive by service user to complete a function previously invoked by an indication

Confirm: Primitive by service provider to complete a function previously invoked by a request primitive

At the transmitting machine, each layer adds a header (a header may not be added at the physical layer) and passes the header to the adjacent layer through the primitive parameters. The header and the data form a protocol data unit (PDU) at each layer, which becomes bigger at each successive lower layer.

The data are transported across the communications media to a receiving station. Here, the data are transferred from the lower layers to the upper layers through the parameters in the primitives. Each succeeding upper layer receives the PDU through the primitive and "strips away" the header pertaining to that layer.

The most important aspect of these operations is that the header created by the transmitting peer layer is used by the receiving peer layer to invoke specific operations according to the rules of the OSI Model.

In later discussions, these concepts are used to show how all the layers operate.

Service definitions

One part of the X.200 Series defines the service definitions between adjacent layers. The abbreviated titles are as follows (see Figure 6.3):

- X.211: Physical Service Definitions

- X.212: Data Link Service Definitions

- X.213: Network Layer Service Definitions

- X.214: Transport Layer Service Definitions

- X.215: Service Layer Service Definitions

- X.216: Presentation Layer Service Definitions

- X.217: Application Layer Service Definitions (Association Control Service Element Definitions)

- X.218: Application Layer Service Definitions (Reliable Transfer Service Element Definitions)

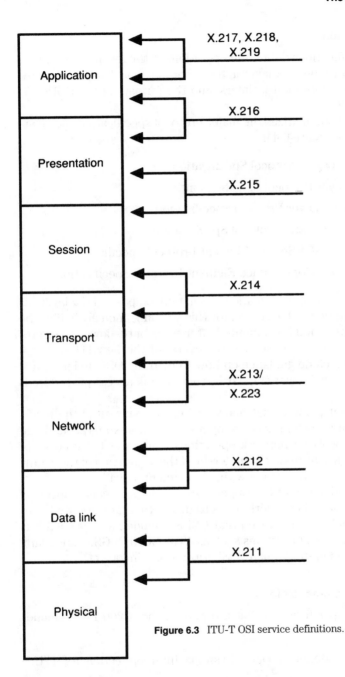

Figure 6.3 ITU-T OSI service definitions.

- X.219: Applications Layer Service Definitions (Remote Operations Service Element Service Definitions)
- X.223: Use of X.25 to provide the OSI connection-mode service for the ITU-T applications

Protocol specifications

The protocol specifications define the operations within each layer and the communications with the peer layer in the other machine. They contain the rules for the use of state tables, timers, and the formats of the PDUs discussed in chapter 2.

The following list summarizes the major protocol specifications using abbreviated titles (see Figure 6.4):

- X.224: Transport Layer Protocol Specification
- X.225: Session Layer Protocol Specification
- X.226: Presentation Layer Protocol Specification
- X.227: Association Control Protocol Specification
- X.228: Reliable Transfer Service Element Protocol Specification
- X.229: Remote Operations Service Element Protocol Specification

The ITU-T OSI protocol specifications do not encompass all the layers as do the service definitions. One reason for this seeming anomaly is that the OSI model was written and implemented after some of the lower-layer protocols were well in place. Consequently, several of the new ITU-T X.200 Recommendations include guidance on how to map the OSI Model service definitions and the lower-layer standards, such as link access procedure B (LAPB) and X.25.

Another important point is that many OSI critics have cited the inefficiency of the OSI approach in network operations. Interestingly, the X.200 Series cite *no* protocols for network operations (the lower three layers of OSI)! This writer is hard-pressed to understand the arguments of these critics. Could it be that they do not know this fundamental fact?

On the other hand, some of the upper-layer X.200 protocols (especially the session layer) have been fairly criticized, for they are not well-conceived. But it should be emphasized that OSI is a model, and as such it is very successful. It is used in systems such as ATM, SONET, GSM, and many others. To call OSI a failure, as many OSI critics do, is incorrect.

Other ITU-T OSI recommendations

Several other OSI specifications that are part of the X.200 Recommendations:

- X.200: Reference Model of Open Systems Interconnection for ITU-T Applications
- X.207: OSI application layer structure
- X.208: Specification for Abstract Syntax Notation One (ASN.1)

X.200 - OSI architecture
X.210 - OSI definition conventions
X.220 - Protocol specifications
X.290 - OSI conformance testing

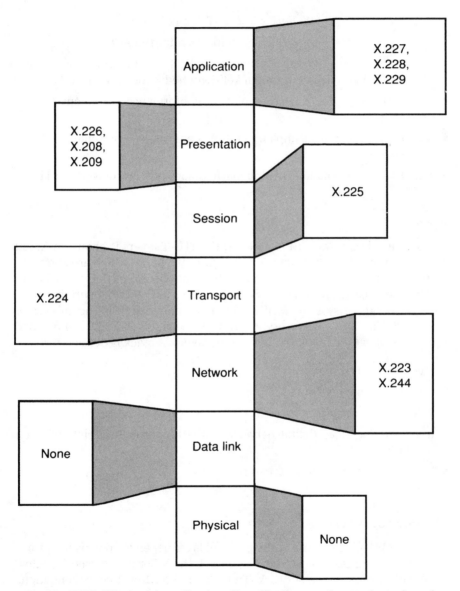

Figure 6.4 ITU-T OSI protocol specifications. Note: The absence of protocols at a layer does not mean the layer contains no protocols. Rather, it means other ITU-T Recommendations are used to fulfill the functions of the layer, such as other X Series, or protocols in the V, I, and Q Series.

- X.209: Specification for Basic Encoding Rules for Abstract Syntax Notation One
- X.210: Open Systems Interconnection Layer Service Definition Conventions
- X.220: Use of the X.200 Series in ITU-T Applications
- X.237: Connectionless protocol specification for the ACSE
- X.244: Procedure for the Exchange of Protocol Identification During Virtual Call Establishment on Packet Switched Public Data Networks
- X.248: RTSE protocol implementation conformance statement (PICS) proforma
- X.249: ROSE protocol implementation conformance statement (PICS) proforma
- X.290: OSI Conformance Testing Methodology and Framework for ITU-T

X.220

The relationship of the OSI Model and the ITU-T layers is shown in X.220 with a diagram. This diagram is divided into two parts, which are shown in Figures 6.5 and 6.6. Figure 6.5 shows the relationships of the upper layers and Figure 6.6, the lower layers. Several entries in Figure 6.4 pertain to the message handling system (MHS). This recommendation is the subject of chapter 8. The X.519 Directory is covered in chapter 9. X.220 defines five protocol suites (stacks) at the lower four layers. A stack is defined for:

- Packet-switched networks
- Circuit-switched networks
- Telephone networks
- Integrated services digital networks (ISDNs) through circuit-switched networks
- ISDNs through packet-switched networks

The Physical Layer

The use of ITU-T standards at the physical layer depends upon which protocol stack is used. As indicated in Figure 6.4, the V Series are used for interfaces with the telephone network. The X.21 and X.21 bis Recommendations are employed with packet-switched and circuit-switched data networks. The ISDN interfaces are provided through the I Series Recommendations for digital networks (I.430 or I.431). The X.21 and X.21 bis Recommendations are discussed in chapter 4 of this book as are other X Series Recommendations appropriate to the physical layer. McGraw-Hill offers separate books by this author on the V and I Series Recommendations.

Like other layers in the OSI Model, the physical layer services are obtained with the invocation of primitives to the physical layer. The ITU-T specifies these primitives in X.211. They are quite simple and are concerned principally with the physical activation and deactivation of circuits and the transfer of data across the physical interface. The physical layer

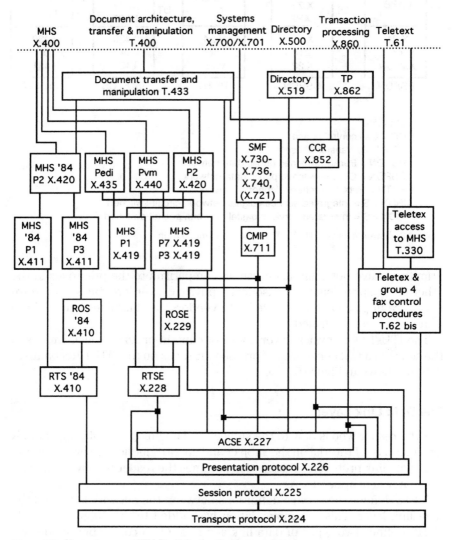

Figure 6.5   Upper layers of ITU-T's X Series Recommendations.

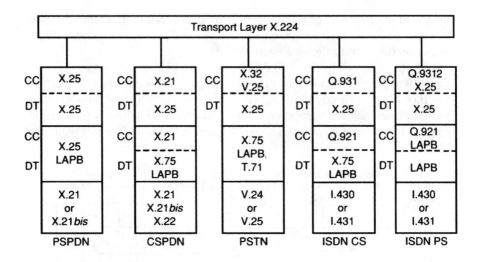

Figure 6.6   Lower layers of ITU-T's X Series Recommendations.

primitives have not seen much use, in comparison to the primitives in the other layers, principally because many physical layer interfaces and protocols were in place and embedded into vendors' product lines before these primitives were published.

The ITU-T OSI physical layer does not contain protocol specifications in the sense of a OSI conventional protocol. It relies on the ITU-T Recommendations shown in Figure 6.5.

The Data Link Layer

For the reader who is new to the subject of data link protocols, appendix A provides a brief tutorial on the subject and an introduction to several prominent data link protocols. This section assumes the reader is conversant with the material in appendix A.

X.212 defines the service definitions between the network layer and the data link layer. This recommendation was added to the 1988 Blue Book. X.212 defines two types of data link services: (1) a connection-mode service and (2) a connectionless-mode service.

The connection-mode service establishes a session between two data link control (DLC) users in order to exchange data link service data units. Inherent in the connection-mode data link service are agreements on certain qualities of service, restrictions on the length of the DLC PDU, the ability to control the flow of the data link user, and the ability to provide a disruptive and possibly destructive release of the data link session.

The connection-mode service of X.212 places strict rules on how these primitives can be issued and which states are allowed to be created from the primitives, as well as which states the primitives can create.

As stated earlier, X.212 also provides for connectionless-mode data link service. As we have learned earlier in this book (chapter 1), a connectionless service is so named because there is no connection established between the data link service users. Therefore, the users do not maintain information about each other. Each transmission is treated as a completely separate event and is not related to previous or succeeding transmissions.

Notwithstanding, the X.212 connectionless-mode service still provides for services such as quality of service (QOS) functions, addressing, and some of the other features found in the connection-mode service. Again, the principal difference between connectionless and connection-mode service is (for connectionless service) the absence of any connection setup prior to the transmission of data.

The ITU-T does not publish an X.200 protocol specification for the data link layer. However, in other recommendations, it defines link layer protocols based on the international standard high-level data link control (HDLC). These protocols are LAPB, which is designed for X.25 interfaces, and Q.921 which is also known as link access procedure for the D channel (LAPD), designed for use on ISDN D channels. Options are also available for the use of another variation of LAPB, called X.75 LAPB. This protocol is almost identical to LAPB, with the exception of the addition of more diagnostics in the frame reject frame (FRMR). The ITU-T uses the HDLC standard in several of its other specifications. For example, a subset of HDLC is used in the popular V.42 and V.42 bis Recommendations, called LAPM (for link access procedure for modems).

An explanation of these protocols can be found in *Data Networks: Concepts, Theory and Practice* (Uyless Black, Prentice Hall, 1989), and *X.25 and Related Protocols* (Uyless Black, The IEEE Computer Society Press, 1991).

The Network Layer

The ITU-T OSI network layer is responsible for providing the transfer of data (more commonly known as packets) between the network service users. Ideally, the users do not know about the specific characteristics of the underlying services. For example, the network service users are unaware of the operations of the underlying network (referred to as a subnetwork or subnetworks). Indeed, if more than one subnetwork is involved in the end-to-end relay, the subnetworks might be dissimilar (heterogeneous); yet, the users are not aware of these dissimilarities.

The network service is also responsible for providing routing and relaying functions between the two users, although ITU-T does not define the routing algorithms (that is a matter for each network to decide). The network service is not concerned with the format, syntax, semantics, or the content of the data it transfers. Consequently, it is not concerned with the upper layers of the OSI Model.

Several of the ITU-T network layer protocols were developed before the OSI Model was completed. As with the lower two layers, specifications have been added to provide guidance on how to use these protocols with the OSI primitives.

X.213 quality of service values for the network layer

Several of the QOS parameters for the transport layer are mapped directly to the complementary quality of services parameters for the network layer. Indeed, these QOS parameters can "begin" at the application layer and can be passed down to the network layer.

The performance aspects of the QOS parameters are shown in Table 6.1a. Table 6.1b shows the nonperformance-oriented parameters.

TABLE 6.1a QOS Performance Parameters

Phase of connection	Performance criteria	
	Speed	Accuracy and reliability
NC establishment	NC establishment delay	NC establishment failure probability (includes misconnection refusal)
Data transfer	Throughput transit delay	Residual error rate NC resilience transfer failure probability
NC release	NC release delay	NC release failure probability

TABLE 6.1b QOS Nonperformance Parameters

NC protection
NC priority
Maximum acceptable cost

The QOS parameters in Table 6.1 are used to invoke the following network services for the network connection between two endpoints. Once the connection is established, the network service users have the same knowledge of the QOS. This pertains to a connection through one network or multiple subnetworks. These QOS parameters are used by many networks today.

NC establishment delay is the maximum acceptable delay between the issuance of the N-CONNECT request primitive and the associated N-CONNECT confirm primitive. This delay also includes the delay encountered at the remote user. Simply stated, it places a time limit on how long a user will wait to get a connection.

NC establishment failure probability is the ratio of total connection failures to total connection attempts. The failure could occur for a number of reasons, such as remote end refusal, expiration of an establishment delay timer, etc. It is a useful parameter for ascertaining the quality of the network service.

Throughput is measured for each direction of transfer through the network. Many private and public networks allow a user to "negotiate" throughput for each connection. Typically, the provision for a higher throughput rate will increase the cost to the user. Consequently, this QOS parameter allows a user to make cost performance trade-offs for each network session. Given a sequence of n data units (DUs) with $n72$, throughput is defined as the smaller of:

1. The number of user data octets contained in the last $n1$ DUs divided by the time between the first and last N-DATA requests in the sequence.

2. The number of user data octets contained in the last $n1$ DUs divided by the time between the first and last N-DATA indications in the sequence.

Transit delay is the time between the N-DATA request and its associated N-DATA indication. The time includes only successfully transferred DUs. Many public networks provide this feature as an option. It is increasingly important as users become more conscious of network performance and response time.

Residual error rate is the ratio of total number of incorrect, lost, or duplicate network service DUs to the total number of DUs transferred. It is defined as:

$$\text{RER} = n(e) + n(l) + N(x)/\text{tsp N}$$

where RER = residual error rate, $n(e)$ = incorrect network service DUs, $n(l)$ = lost DUs, $N(x)$ = duplicate DUs, and N = total DUs transferred. This QOS parameter is useful when a user wishes to obtain a quantitative evaluation of the network service reliability.

Transfer failure probability is the ratio of total transfer failures to total transfers. The ratio is calculated on an individual network connection. A failure is defined by an observed performance that is worse than the specified minimum level and is measured against three other QOS parameters: throughput, transit delay, and residual error rate.

Network connection resilience is defined as the probability of the network service provider (1) invoking a disconnect and (2) invoking a reset. Both actions come as a result of network actions and not as a user request. This parameter is quite important in evaluating the quality of a network. Several public networks publish their performance data relating to connection resilience.

Network connection release delay is the maximum acceptable delay between a user disconnect request and the receipt of a disconnect indication, that is, when the user is able to initiate a new network connection request. This is an important consideration for users who need fast connections after "logging off" from a previous network session.

Network connection release failure probability is defined as the probability that the user will not be able to initiate a new connection with a specified maximum release delay. It is the ratio of total network connection release failures to total release attempts.

Network connection protection is a qualitative QOS parameter to determine the extent to which the network service provider attempts unauthorized use of data. This parameter, if used by a public network, is not often revealed to a user.

Network connection priority is defined as (1) the order in which the connections have their QOS degraded and (2) the order in which connections are broken. Typically, once a user obtains a network connection, it remains stable for the duration of the session.

Maximum acceptable cost, as its name implies, specifies the maximum acceptable cost to the user for a network connection. The costs are specified by the network in relative or absolute terms.

In summary, X.213 serves as a model for defining network interfaces and network services. The X.213 primitives and QOS parameters are now used by a number of network vendors.

Relationship of the X.200 Recommendations to other ITU-T network layer recommendations

The ITU-T X Series contains no "pure" OSI network layer protocol specification. However, X.25 is commonly used at the network layer for connection-mode packet-based network service, the ISDN I Recommendations are used for digital networks, and X.21 is used for a connection-mode circuit-based network service.

X.25 was developed several years before the OSI Recommendations on the network layer were published. Therefore, the ITU-T has made efforts to bring the X.25 specification within the OSI umbrella and publishes X.223 to define the mapping between X.25 and the primitives of X.213.

The arrangements for using X.223 for call management are shown in Table 6.2, which shows the X.213 connection primitives and their mapped counterparts in the X.25 packet procedures. In addition, the table also shows the X.213 primitive parameters and how they are mapped to the fields in X.25 connection management packets. The X.213 QOS parameters are mapped into X.25 packet level procedures, principally through the use of X.25 facilities. Table 6.2 also shows the mapping relationship of the QOS subparameters and how they relate to the X.25 packets and the X.25 facilities.

The X.25 Data packets are created from the X.213 data primitives. These are the N-DATA request primitive and the N-DATA indication primitive. These primitives carry two parameters which are mapped into fields of the X.25 packet. First, the NS-user-data parameter is used to create the X.25 user data field and perhaps the M bit. Secondly, the X.213 confirmation request primitive is mapped into the X.25 D bit and the P(S) field.

As might be expected, the X.25 Interrupt packet is created from the X.213 expedited data request primitives, and the X.213 reset primitives are mapped into the X.25 Reset packets. Finally, the disconnect primitives are mapped into the X.25 Clear packets, which are also shown in the table.

The Transport Layer

The ITU-T transport layer was developed at about the same time as the OSI Model was developed. Therefore, it is aligned with the concepts and terminology found in the X.200 specifications.

Introduction

The transport layer is quite important to user applications that require assurance that all PDUs have been delivered safely to the destination. Perhaps the best approach is to think of the transport layer as a security blanket. It attempts to take care of the data, regardless of the goings-on in the underlying network(s) (or the lower layers of OSI).

The transport layer performs many other functions. One of its principal jobs is to shield the upper layers from the details of the lower layers' operations. Indeed, the user may be completely unaware of the physical network(s) that are supporting the user's activities, because the transport layer is providing the transparent interface between the user and the network(s). Ideally, the transport service relieves the upper layers of concern about how to obtain a needed level of network service.

TABLE 6.2   X.213 Primitives and X.25 Packets (X.223)

X.213 primitives	X.25 packets
N-CONNECT request	Call Request
N-CONNECT indication	Incoming Call
N-CONNECT response	Call Accepted
N-CONNECT confirm	Call Connected
N-DISCONNECT request	Clear Request
N-DISCONNECT indication	Clear Indication, Restart Indication, or Clear Request
N-DATA request	Data
N-DATA indication	Data
N-EXPEDITED DATA request	Interrupt
N-EXPEDITED DATA indication	Interrupt
N-RESET request	Reset Request
N-RESET indication	Reset Indication, Reset Request
N-RESET response	
N-RESET confirm	

Primitive parameters	Fields in the packets
Called Address	Called DTE address field or Called address extension facility
Calling Address	Calling DTE address field or Calling DTE address extension facility
Responding Address	Called DTE address field or Called DTE address extension facility
Receipt Confirmation Selection	General format identifier (GFI)
Expedited Data Selection	Expedited data negotiation facility
QOS Parameters	These X.25 facilities: throughput class negotiation, minimum throughput class negotiation, transit delay selection and indication, and end-to-end transit delay
Originator and Reason	Cause code and diagnostic code
NS-User Data	User data field in a Fast Select packet
NS-User Data	User data field in a Data packet
Confirmation Request	D bit with the P(S)

These ideas are shown in Figure 6.7. First, the transport layer resides outside the networks and therefore is not subject to any network problems. Second, it resides between the upper layers of the user application and the lower layers of the networks and therefore keeps the network operations transparent to the upper layers.

As the reader might expect, the transport layer service definition uses primitives to specify what services are to be provided through the network(s). The parameters associated with each primitive action provide the specific actions and events to be provided by the network. During the connection establishment phase, the characteristics of the connection are negotiated between the end users and the transport layer. The primitives and the parameters of the primitives are used to support the negotiation. It is possible that a connection will not be made if the network or an end user cannot provide the requested QOS. Moreover, the requested services may be negotiated down to a lower level.

Types of networks and classes of protocol

The ITU-T OSI transport layer protocol specifications are contained in X.224. They are based on two key concepts: (1) types of networks and (2) classes of protocol.

The quality of network services rests on the types of networks available to the end user and the transport layer. Three types of networks are defined by the ITU-T recommendations.

A *type A* network provides acceptable residual error rates and acceptable rates of signalled failures (acceptable quality). Packets are assumed not to be lost. The transport layer need not provide recovery or resequencing services. In other words, it is an ideal network.

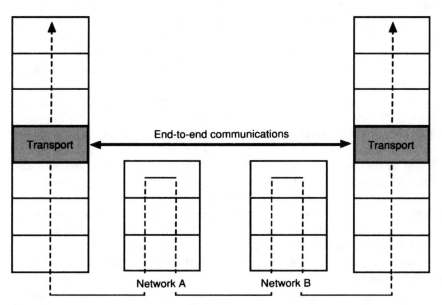

Figure 6.7 The transport layer peer entities.

A *type B* network provides for acceptable residual error rates, but for an unacceptable rate for signalled failures (unacceptable signalled errors). The transport layer must be able to recover from errors.

A *type C* network connection provides a residual error rate not acceptable to the user (unreliable). The transport layer must recover from failures and resequence packets. A type C network could include some local area networks, mobile networks, and datagram networks.

The transport layer is responsible for selecting an appropriate protocol to support the QOS parameters established by the user. Since the transport layer knows the characteristics of a network (types A, B, or C), the layer can choose five classes of protocol procedures to support the QOS request from the user. The protocol classes are:

- Class 0: Simple class
- Class 1: Basic error recovery class
- Class 2: Multiplexing class
- Class 3: Error recovery class
- Class 4: Error detection and recovery class

The *class 0* protocol provides for a very simple transport connection establishment to support a type A network. Class 0 provides for a connection-oriented support, both during the network connection and the release phases. It does not provide for any support of user transfer data during connection establishment. This protocol is able to detect and signal protocol errors, but it does not recover from errors. If the network layer signals an error to the transport layer, the transport layer releases the connection to its network layer. In so doing, the end user is informed about the disconnection. The ITU-T developed this class for its teletex standard (ITU-T T.70).

The *class 1* protocol is associated with networks such as an X.25 packet network. In contrast to class 0, the transport PDUs are numbered. The class 1 protocol provides for the segmenting of data, if necessary, and the retention of all data and acknowledgments. It allows resynchronization of the session in the event of an X.25 Reset or Restart packet. The protocol also supports expedited data transfer. It responds to disconnect requests and protocol errors. It is also responsible for resynchronization and performing reassignments in the event of a network failure (i.e., an X.25 Reset packet).

It is possible to transmit user data in a class 1 connection request. Also, each DU is sequenced to aid in ACKs, NAKs, and error recovery. The ACKs release the copies of the DUs at the transmitting sites. Class 1 also provides for either user acknowledgment or network receipt acknowledgment.

The *class 2* protocol is an enhancement to class 0 that allows the multiplexing of several transport connections into a single network session. It also provides for explicit flow control to prevent congestion from occurring at the end user sites. Class 2 provides no error detection or recovery. If a Reset or Clear packet is detected, this protocol disconnects the session and the user is so informed. The class 2 protocol is designed to be used over very reliable type A networks. User data can be transmitted in the connection request DU.

The *class 3* protocol provides for the services included in the class 2 structure. It also provides recovery from a network failure, without requiring the notification of the user. The user data are retained until the receiving transport layer sends back a positive acknowledgment of the data. This class has a very useful mechanism to retransmit data. The packets in transit through a network are given a maximum "lifetime" through timers. All data requiring a response are timed. If the timer expires before an acknowledgment is received, retransmission or other recovery procedures can be invoked. The class 3 protocol assumes a type B network service.

The *class 4* protocol includes the flow control functions of classes 2 and 3 and assumes that practically anything can go wrong in the network. Like class 3, expedited data are allowed and the ACKs are sequenced. This protocol allows for "frozen" references (a reference is a connection identifier). Upon a connection release, the corresponding reference cannot be reused, since the network layer could still be processing late-arriving data associated with the references. The class 4 protocol is designed to work with type C networks.

In summary, the protocol classes and the associated network types are:

Class	Network type
0	A
1	B
2	A
3	B
4	C

The Session Layer

The session layer is so named because it manages the user application-to-application sessions. The lower layers are not concerned with the users' applications dialogues, and while the data link, network, and transport layers all have certain data recovery functions, they may lose data. It is the task of the session layer to ensure that the DUs discarded at the lower layers are recovered. Moreover, it is the first layer we have examined in the OSI Model that permits a "graceful close": ensuring all data units have been received safely before the applications are allowed to close their session.

Most session layer systems provide the following services:

- Coordinating the exchange of data between applications by logically connecting and releasing sessions between the applications (also called dialogues)

- Providing synchronization points (by the exchange of sync point "tokens") to structure the exchange of data

- Imposing a structure on the user application interactions

- Providing a convention to take turns in exchanging data by exchanging a data token to determine which application can send data

Synchronization (sync) points are an important part of the session layer. They are used to coordinate the exchange of data between session service users. Synchronization services are like checkpoint restarts in a file transfer or transaction transfer operation. They allow the users to (1) define and isolate points in an ongoing data exchange and (2) if necessary, back up to a point for recovery purposes.

Two types of sync points are available: (1) minor sync points and (2) major sync points. As Figure 6.8 illustrates, the sync points are used in conjunction with the following dialogue units and activities:

Major sync point. Structures the exchange of data into a series of dialogue units. Each major sync point must be confirmed and the user is limited to specific services until a confirmation is received. The sending user can send no more data until the receiving user acknowledges a major sync point. A major sync point allows related units of work to be clustered together.

Dialogue unit. An atomic action in which all communications within it are separated from any previous or succeeding communications. A major sync point delineates the beginning and ending of a dialogue unit.

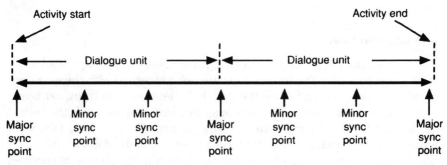

Figure 6.8 Session layer operations.

Minor sync point. Structures exchange of data within a dialogue unit. This is more flexible than a major sync point. For example, each sync point may or may not be confirmed and it is possible to resynchronize to any minor sync point within the dialogue unit. A confirmation confirms all previous minor sync points.

Activity. Consists of one or more dialogue units. An activity is a logical set of related tasks, for example, the transferral of a file with related records. A useful feature of the dialogue unit is that it can be interrupted and later resumed. Each activity is completely independent of any other activity.

The OSI Model does not specify what is to be managed with sync points and dialogue units. As examples, a file transfer system could be managed with each complete file considered as a dialogue unit and delineated with major sync points, or an inquiry response system could be managed with each individual transaction delineated with a major sync point.

Session layer organization

The session layer is organized around the idea of *functional units.* The functional unit concept is used to define and group related services offered by the layer. Table 6.3 describes these units.

TABLE 6.3   Session Services Functional Units.

Kernel	Provides five nonnegotiable services, a minimum subset of the session layer services.
Half-duplex	Alternate two-way transmission of data between SS users. Data sent by owner of data token.
Duplex	Simultaneous two-way transmission of data between SS users.
Typed data	Transfer of data with no token restrictions.
Exceptions	Reporting of exceptional situations by either SS user or SS provider.
Negotiated release	Releasing a session through orderly (normal) measures or by passing tokens.
Minor synchronize	Invoking a minor sync point.
Major synchronize	Invoking a major sync point.
Resynchronize	Reestablish communications and reset connection.
Expedited data	Transferring data that are free from token and flow control restarts.
Activity management	Providing several functions within an activity.
Capability data exchange	Exchanging a limited amount of data while not operating within an activity.

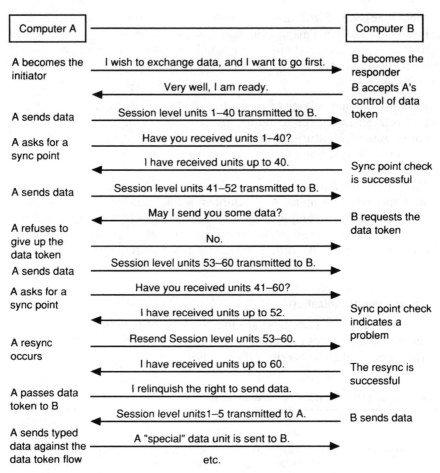

Figure 6.9 Session layer operations.

Figure 6.9 shows an example of the use of several session layer operations. The user at computer A becomes the initiator of the session and reserves the right to send data by keeping the data token. The user at computer B (the responder) accepts this structure for the session. User A then sends a number of DUs and periodically asks for a sync point to make certain the data are received correctly at B. If the sync point check is not successful, the session layer entities initiate a resynchronization. Further down in the figure, B asks for the data token but is refused because A has more data to transfer. Later, A relinquishes the data token, and B sends data. Afterward, A issues a typed DU (which allows a small amount of data to be sent against the data flow).

The Presentation Layer

The presentation layer performs the services of data structure description and representation. It is not concerned with the semantics of the data, but with preserving the syntax of the data. It is often used to resolve the internal data representation differences between computers. For example, some computers store numbers in one's complement and other computers store the same numbers in two's complement. This layer can also accept a data type (such as ASCII, Boolean, integer, etc.) and negotiate the conversion of the type to another type.

The presentation layer is concerned with (1) the syntax of the data of the sending application, (2) the syntax of the data of the receiving application, and (3) the data syntax used between the presentation entities that support the sending and receiving applications.

The latter service is called a *transfer syntax* and it is negotiated between the presentation entities. Each entity chooses a syntax and then attempts to negotiate the use of this syntax with the other presentation layer entity. Therefore, the two presentation entities must agree on a transfer syntax before data can be exchanged. Moreover, it may be necessary for the presentation layer to transform the data in order for the two users to communicate.

Example of syntax negotiation

We just learned that a user relies on the presentation layer for two important operations: (1) an abstract syntax agreement between the OSI entities and (2) agreement on a transfer syntax between the two communicating entities. Figure 6.10 provides an example of how an application process uses the services of the presentation layer to achieve these two services. As shown in the figure, the application entity sends a P-CONNECT request primitive (defined in X.216) to the presentation layer. Some of the parameters in this primitive contain the abstract syntax identifications (ABSYNA, ABSYNB). The presentation entity receives this information and decides if it can support these syntaxes with the use of a transfer syntax. It then encodes a connection request presentation protocol data unit (PPDU) to send to the presentation entity in the other machine. Included in the PDU are the proposed abstract syntax(es) (ABSYNA, AYSYNB) as well as the proposed transfer syntax(es) (TRNSYN1, TRNSYN2, TRNSYN3) it is able to support for each abstract syntax presented to it from the user application entity. The receiving presentation entity receives the PDU. It then issues a P-CONNECT indication primitive to the application layer.

The application entity sends back a P-CONNECT response primitive. This primitive contains the names of the abstract syntaxes it is able to use during the process (only ABSYNB). It cannot negotiate new abstract syn-

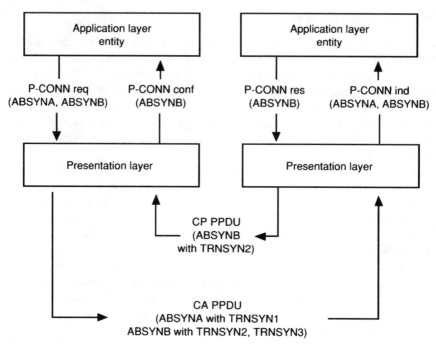

Figure 6.10   Presentation service negotiation.

taxes at this time. It must use at least one of the abstract syntaxes it received in the P-CONNECT indication primitive or refuse the connection.

The remote presentation entity receives this primitive and maps a PDU from the parameters. It indicates in the PDU the transfer syntaxes that are to be used during the communications process (ABSYNB with TRNSYN2).

Finally, the local presentation entity receives the PDU and examines the proposed transfer syntaxes. If it accepts the proposals, it maps these fields to the P-CONNECT confirm primitive and sends it to the local application entity.

In summary, these operations resulted in the negotiation of the abstract syntax and the transfer syntax between the pair of communication entities. The final negotiated agreement is called a *presentation context*. Furthermore, the current presentation context containing the specific abstract syntax(es) and transfer syntax(es) is called the *defined context set*.

The X.208 and X.209 Recommendations

The ITU-T also specifies X.208 and X.209 for the presentation layer. X.208 specifies the ASN.1 language and X.209 specifies the basic encoding rules (BER) for ASN.1. ASN.1 permits a data element to be described by (1) defining its type (integer, etc.) and (2) optionally assigning a user-friendly name to it. In the 1988 Blue Books, the X.208 specification was aligned with ISO 8824, plus ISO 8824, Addendum 1 (except 8824 does not define some conventions on describing encrypted structures). X.209 is aligned with ISO 8825 plus ISO 8825, Addendum 1.

ASN.1 is used extensively at the OSI upper layers, but it need not be restricted to these layers. Notwithstanding, OSI now requires that all data exchanges between the application and presentation layers must be described in abstract syntax.

Figure 6.11 shows the structure of ASN.1. Each piece of information exchanged between users has a *type* and a *value*. The type is a *class* of information, such as integer, Boolean, octet, etc.

To distinguish the different class types, a structure of values (for example, a database record) or a simple element (for example, a field within the database record) can have a tag attached that identifies the type with a *number*. For example, a tag for a point-of-sale transaction could be the number 22. This number is used to identify the record and inform the receiver about the nature of its contents.

ASN.1 defines four types of *classes;* each tag is identified by its class and its number (as in our example, 22). The type classes are defined as:

Universal. Application-independent types

Applicationwide. Types that are specific to an application and used in other standards (X.400, X.500, etc.)

Context-specific. Types that are specific to an application but are limited to a set within an application

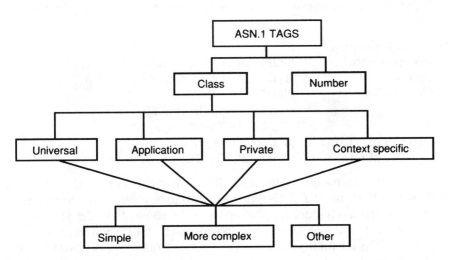

Figure 6.11   The X.208 ASN.1 scheme.

Private-use. Reserved for private use and not defined in these standards

Several tags are used for the universal assignment and the numbers associated with these tags are reserved. For example, a bit string type has Universal 3 reserved for it.

Finally, the encoding may consist of only one type, in which case it is classified as *simple* type. On the other hand, it is possible to represent more than one type in a *more complex* type, such as a sequence of types.

The *other* box in Figure 6.11 allows us to do anything we want.

Example of ASN.1 notations

The best way to explain these concepts is through an example. This example is a PDU used with the reliable transfer service element (RTSE, discussed later in this chapter).

In order to be able to read this example, it is useful to know the ASN.1 rules for using uppercase and lowercase letters in names. The rules are:

- All type names begin with an uppercase letter.
- All module names (explained shortly) begin with an uppercase letter.
- Names of built-in types and other reserved words are all uppercase letters.
- All other identifiers begin with a lowercase letter.
- Comments can be entered to explain the ASN.1 notations. Comments are preceded by two hyphens (--).

ASN.1 is a free-format language. The programmer can terminate the code within the language anywhere on a line, because it is merely interpreted as a blank (with the exception of comments). Our example is coded in ASN.1 as follows:

```
RTORQapdu ::=        SET {
checkpointSize       [0] IMPLICIT INTEGER DEFAULT 1,
windowSize           [1] IMPLICIT INTEGER DEFAULT 3,
dialogueMode         [2] IMPLICIT INTEGER
                         {monologue(0),twa(1)}
                         DEFAULT monologue,
connectionDATARQ     [3] ConnectionData,
applicationProtocol  [4] IMPLICIT INTEGER OPTIONAL
```

The notation ::= means "defined as." The type name *RTORQapdu* is described with the type *SET*. This type is defined to be a series of other types, bounded by the [and] brackets. The order of the elements in the *SET* does not matter; any of the elements can precede each other when they are encoded. A *SET* is useful to define a known number of elements whose order is not important.

Several of the fields are named and typed as integer with the ASN.1 built-in type *INTEGER*.

Part of the notation in this example is: *checkpointSize* [0] *IMPLICIT INTEGER DEFAULT 1,*. This definition assigns a new tag value to the built-in type of *INTEGER*. It causes the original type value to be replaced with the tag whose class is private and whose value is 0. (We might wish to code the user data element as private because it is used in a company-specific protocol.) In the actual transfer syntax between two computers, the *INTEGER* tag would be omitted. This feature is a very useful option of ASN.1 because the transmitted PDU would not have to carry the new and old tag, just the new tag. Consequently, fewer bits are transmitted.

ASN.1 allows fields to be defined as *DEFAULT*. This reserved word is followed by the value to be used as a default value if a value is not furnished. In this example, the default value for *windowSize* is *3*.

The user-friendly names for the values must begin with a lowercase letter (*checkpointSize, applicationProtocol*, etc.). ASN.1 does not act on these names, only names with uppercase letters. To enhance the readability of these identifiers, uppercase letters are usually inserted for each new meaningful, user-friendly word embedded in the identifier.

Notice that all entries have been coded with a type (*INTEGER*) except the line with a [3]. Since this line has the name *ConnectionData* (which begins in uppercase), it must be defined with a type somewhere in the ASN.1 code. The code to define the type is not shown in this example, and under real conditions, the reader would have to go to another part of the ASN.1 printout to learn about this entry.

Like any computer-based language, ASN.1 takes time to learn. It also takes practice. However, once you have used it a number of times, some of the rules become fairly easy to apply.

Basic encoding rules

X.208 ASN.1 by itself is not particularly useful during the transfer of data between machines because the language does not specify the length of the data. Therefore, X.209 is employed to provide rules on how to describe the length of each data element. The data element consists of three components, which appear in the following order: *Type Length Value* (TLV).

The type value (also called an identifier) distinguishes one type from another and specifies how the contents are interpreted. The length specifies the length of the actual value. The value (also called contents) contains the information of the element.

The data element that is transmitted on the communications channel can consist of a single TLV or a series of data elements, described as multiple TLVs.

The type field is coded as shown in Figure 6.12. Bits 8 and 7 identify the four type classes with the following bit assignments:

Application-wide Universal	00
Application-wide	01
Context-specific	10
Private-use	11

Bit 6 identifies the forms (F) of the data element. Two forms are possible. A primitive (simple) element (bit 6 = 0) has no further internal structure of data elements. A constructor (complex) element (bit 6 = 1) is recursively defined in that it contains a series of data elements, such as a SEQUENCE or SET of types. The remaining 5 bits (5 through 1) distinguish one data type from another class. For example, the field may distinguish Boolean from integer. If the system requires more than 5 bits, this field is coded to allow an extension to additional octets.

The length (L) specifies the length of the value. It may take one of three forms: short, long, or indefinite to define an element of varying lengths. The value field contains the actual information. It is described in multiples of 8 bits and is variable in length. The contents are interpreted based on the coding of the type and length fields. Therefore, the contents are interpreted as bit strings, octet strings, etc.

Figure 6.13 shows how the example of the RTORG PDU might appear on a communications channel. The contents of the type fields are determined by the ASN.1 code examined earlier. The contents of the length and value fields depend upon the actual data that are to be transferred between the machines.

This brief explanation of the presentation layer does not do justice to all its functions and capabilities. However, it should be sufficient to give the reader an idea of how it operates.

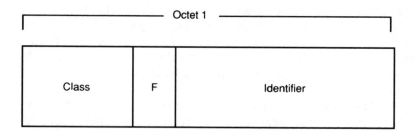

Class = class of the data type
F = form of the element
Identifier = identification of the element

Figure 6.12 The type field.

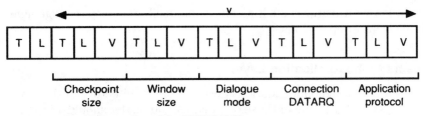

Figure 6.13   The TLV encoding of the RTORG PDU.

The Application Layer

Introduction

It is quite possible to devote an entire book to the application level. Its functions are broad and encompass protocols as diverse as office document systems, electronic mail, file transfer, and terminal control. The 1984 Red Books were quite "spotty" in their coverage of the applications layer, but the 1988 Blue Books have filled many of the gaps.

We examine the architecture of the application layer with the ITU-T X.200 Recommendations listed below. Please refer to Figure 6.5 during this discussion. This figure also shows that other ITU-T Recommendations, such as X.400 and X.500, are related to this layer. However, as separate X Series Recommendations, they are examined in other chapters. The ITU-T application layer architecture is based on:

- X.207: OSI application layer structure
- X.217 and X.227: Association Control Service Element (ACSE) and Association Control Protocol
- X.218 and X.228: Reliable Transfer Service Element (RTSE) and eliable Transfer Protocol
- X.219 and X.229: Remote Operations Service Element (ROSE) and Remote Operations Transfer Protocol
- X.237: Connectionless protocol specification for the ACSE
- X.248: RTSE protocol implementation conformance statement (PICS) proforma
- X.249: ROSE protocol implementation conformance statement (PICS) proforma

The application layer is not intended to include the end user application. That is, the user application does not reside within the application layer but on top of it. Some people consider the name "application layer" to be a misnomer, but it is aptly named because it provides *direct* services to the user application with features such as electronic mail, database management, and file transfer operations.

Aspects of the Application Layer

Several X Series Recommendations describe the architecture of the application layer. X.207 is one of those documents; the reader should also study X.650. X.207 is actually a supplement to X.200 in that it expands and refines several application layer concepts. The application-layer architecture, and its associated naming, addressing, and identification conventions are complex. The reader should consult X.200, X.207, X.650, and X.660 for more information on this subject. These specifications are rather abstract, and will most likely be of interest only to designers. The remainder of this section acquaints you with the major aspects of X.207.

To begin the analysis, the "bundling up" of the various application service elements needed to support user applications is called an application context. The application context is a set of rules that are shared by two application service objects (ASOs). This ASO is merely an active (executing) process in a computer.

The ITU-T uses the term application process type in the following manner: it consists of the computer resources to represent a service, such as E-mail. It is identified with an application process title. In turn, an entity type consists of a specific class of resources, such as X.400 interpersonal messaging system. In order to obtain these services, the software, databases, etc. must be invoked. This process in called an application process invocation (AP-I to n, in this figure), which starts an application entity invocation (AE-1 to n, in Figure 6.14). The AE invocation begins the establishment of the application association (AA).

An AP invocation is identified with an AP invocation identifier, and an AE invocation is identified with an AE invocation identifier. One might question why two identifiers are needed. It is possible to invoke multiple AEs within one AP invocation. Therefore, identifiers are needed for all invocations. Moreover, all these identifiers are optional. Indeed, only one identification is required to set up an application association—an application context name.

Once again, the reader is encouraged to obtain and study the X.200, X.207, X.650, and X.660 Recommendations if more information is needed on this subject. Figure 6.13 provides a list and discription of these identifiers.

X.227 Association Control Service Element

The ACSE provides services to other service elements, such as RTSE, and X.400 service elements. Its main purpose is to support the establishment, maintenance, and termination of *application associations*. An application

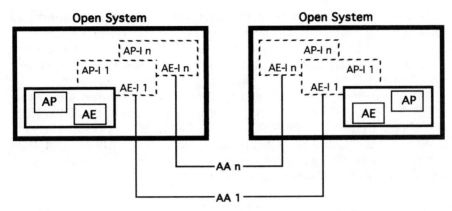

Where:

   AP-I   = Application process invocation

   AP-E   = Application entity invocation

Note:

   1. AP invocation is a run-time instance
      of an application process

   2. AP run-time instance creates the AA
      through the AE invocation

   3. A single AP invocation may be mapped into:
      one or more AE invocations

Figure 6.14   Relationship of AP invocation and AE invocation.

association is a detailed description of a cooperative relationship between two application entities through the use of presentation services within a defined *application context*. An application context is simply the set of service elements and supporting information used on the application association, something like a cataloged procedure in the IBM world.

ACSE's main task is to support associations between applications. In so doing, it performs four services: (1) A-ASSOCIATE, (2) A-RELEASE, (3) A-ABORT, and (4) A-P-ABORT.

The A-ASSOCIATE service sets up the association between the applications. It is responsible for performing the following services:

- Supporting application entity titles (optional)
- Supporting the application context name (required)
- Providing the presentation context (optional)
- Providing a result of the operation (required)

As might be expected, the A-RELEASE service provides an orderly release of the association of the two applications. It can use the session layer orderly release to ensure both applications agree to the release. Otherwise, the association continues. The abort services (A-ABORT, and A-P-ABORT) cause the termination of the application, presentation, and session connections. The A-ABORT is initiated by the user or the association control service. The A-P-ABORT service is initiated by the presentation layer.

The service user is not required to directly use all the features of the ACSE. For example, an electronic message service such as the X.400 MHS may choose to use the A-CONNECT and A-RELEASE services and rely on a service element such as ROSE to make use of the A-ABORT and A-P-ABORT services. How all this fits together is determined by the people who actually establish the application context.

X.237 Connectionless ACSE

The ACSE connectionless mode was added to the ITU-T X Series in 1992. It uses the X.217 service definitions, and adds an additional protocol data unit, which is called the A-UNIT-DATA ACSE PDU. It also uses the P-UNIT-DATA presentation service.

As a connectionless protocol, X.237 is quite simple. Its principal job is to submit a PDU from one AEI to another AEI. It is important to note that X.237 does not describe how the AEIs are created.

Of course, ASN.1 is used to define the contents of the APDU. Most of the fields in the APDU are optional and/or defined by the user. Two fields are required: (a) the application context name, and (b) user information. This PDU is registered under the following naming registration hierarchy (see X.660 for more information on this subject): joint-iso-ccitt association-control(2) module(2) acse(1) version(1).

X.228 Reliable Transfer Service Element

In recent years, the ITU-T has recognized the need to define an application service element to support the reliable transfer of application protocol data units (APDUs) between applications and to ensure that the sender is notified if the delivery cannot be made. Obviously, these services are often applicable to more than one application. Therefore, they have been defined as a common service element, RTSE. In addition to these services the RTSE recovers from communication and end-system failures. RTSE provides seven services: (1) RT-OPEN, (2) RT-TRANSFER, (3) RT-TURN-PLEASE, (4) RT-TURN-GIVE, (5) RT-P-ABORT, (6) RT-U-ABORT, and (7) RT-CLOSE.

An association is established between two application entities (AEs) with the RT-OPEN service. The association-initiator identifies the application context name for normal mode. The service uses the conventional OSI request, indication, response, and confirm primitives because it is a confirmed service. The majority of the parameters in these primitives are passed to ACSE or the presentation layer. After the association is established, the RT-TRANSFER service is invoked to transfer an APDU.

As the reader might expect, the RT-TURN-PLEASE and the RT-TURN-GIVE services enable the peer RTSE-user to request or relinquish the turn. The user cannot transfer an APDU if it does not have the turn. Moreover, the RTSE-user must also have the turn in order to release an application-association.

The RT-P-ABORT service is used by the RTSE-provider to indicate it cannot maintain the application-association. This provider is responsible for taking the following actions to ensure the integrity of the APDU transfer: (1) if it is the APDU sender, it must first issue a negative confirm in an RT-TRANSFER confirm primitive for the nondelivered APDU, and (2) if it is the APDU receiver, it must delete any partially received APDU before it issues the RT-P-ABORT indication primitive.

The RT-CLOSE service is used by the RTSE-user to request the release of the application-association. The requester can only evoke this service if it has possession of the turn.

X.248 RTSE Protocol Implementation Conformance Statement (PICS) Proforma

This specification was released in the fall of 1992. As its name suggests, it defines the procedures for an organization (such as a test lab) to verify that a supplier's RTSE conforms to the X.228 Recommendation. It also explains which other protocols may be affected by an application context in which RTSE is used and provides reference to several sections in X.482, X.483, and X.484.

X.229 Remote Operations Service Element

The remote operations specification was published in 1984 with the X.400 MHS recommendations (X.410). It was recognized it was applicable to general use and has been published in the 1988 Blue Books as *X.229*.

Most applications designers develop interactive applications based on three operations: (1) a transaction from an originator to the recipient, (2) a transaction from the recipient to the originator, and (3) an error report about problems resulting from the transactions. Furthermore, the originator (or invoker) can invoke operations in another system (the performer). Since the systems may be of different types and the communicating parties separated by distance, the name ROSE has been coined to describe these procedures.

ROSE operates by reporting the success or failure of an operation. It defines several ways these reports are sent by the operation performer. Table 6.4 summarizes these operations.

ROSE permits two modes of operation. The *synchronous mode* requires the performer to reply to the invoker before the invoker can invoke another operation, which is a common method to control interactive, transaction-based systems. The *asynchronous* mode allows the invoker to continue invoking operations without requiring a reply from the performer.

The reporting requirements (Table 6.5) are combined with the mode definitions to define five classes of operations. These operations are also summarized in Table 6.5.

ROSE provides for the grouping of operations in the event that the performer notifies the invoker that it must also perform some operation(s). This function is called *linked operations*. The invoker of an operation invokes a parent operation, which is executed by the performer (i.e., the performer of a parent operation). This AE can invoke one or more child operations, which are performed by the invoker, i.e., the performer of child operation(s).

ROSE uses four macros, which are defined in ASN.1. They are:

BIND. Allows the specification of the types of user data values to be exchanged in the establishment of the application association.

UNBIND. Allows the specification of the types of user data values to be exchanged in the release of the application association.

TABLE 6.4   Classification of ROSE Operations

Result of operation	Expected reporting from performer
Success or failure	If successful, return result If a failure, return an error reply
Failure only	If successful, no reply If a failure, return an error reply
Success only	If successful, return result If a failure, no reply
Success or failure	In either case, no reply

TABLE 6.5   ROSE Operation Classes

Class number	Definition
1	Synchronous: Report success (result) or failure (error)
2	Asynchronous: Report success (result) or failure (error)
3	Asynchronous: Report failure (error) only
4	Asynchronous: Report success (result) only
5	Asynchronous: Report nothing

OPERATION. The type notation in this macro allows the specification of an operation and user data to be exchanged for a request and a positive reply. This macro can specify a list of linked child operations if the operation is a parent operation.

ERROR. The type notation in this macro allows the specification of an operation and user data to be exchanged in a negative reply.

ROSE provides for five services:

- RO-INVOKE
- RO-RESULT
- RO-ERROR
- RO-REJECT-U
- RO-REJECT-P

The *RO-INVOKE* service allows the invoker AE to request that an operation be performed by a performer AE. This service begins when the ROSE receives a request from the ROSE user (the invoker) to be performed by the other ROSE user (the performer). This service is a nonconfirmed service.

The *RO-RESULT* service is also a nonconfirmed service. It is used in reply to a previous RO-INVOKE to notify the invoker of the completion of a successful operation.

The *RO-ERROR* service is a nonconfirmed service to reply to a previous RO-INVOKE. Its purpose is to notify the user of an unsuccessful operation.

The *RO-REJECT-U* service is a nonconfirmed service which rejects an RO-INVOKE service. It is also used to reject an RO-RESULT and an RO-ERROR, if necessary.

Finally, the RO-REJECT-P service is used by the ROSE provider to inform the ROSE user about a problem.

X.249 ROSE Protocol Implementation
Conformance Statement (PICS) Proforma

This specification was released in the fall of 1992. As its name suggests, it defines the procedures for an organization (such as a test lab) to verify that a supplier's ROSE conforms to the X.229 Recommendation. It also explains which other protocols may be affected by an application context in which ROSE is used and provides reference to several sections in X.483, X.484, and X.581.

How the Layers Handle Data Integrity Issues

This book has examined several ways the X Series Recommendations allow an end-user to issue signals to request a network to perform certain operations. We are concerned with how each layer:

- Connects users
- Transfers data between users
- Disconnects users

Three important operations for disconnect services are the *clear, reset*, and *restart*.

The ITU-T recommendations vary in how connect, data transfer, and disconnect operations are used in a particular recommendation and layer. For example, their use and effect in an X.25-based packet-switched network varies from their use and effect in an X.21-based circuit-switched network. But common threads run among them. Logically enough for network services, they are applied at the network layer of the OSI Model. However, other forms of these three types of operations can be invoked at several of the other layers. We pause briefly to examine these operations because of their relevance to user data integrity.

Physical layer

The issue of connection and data transfer integrity is generally not an issue at the physical layer. This layer acts as the physical medium for the data transmissions and relies on the upper layers for user data integrity.

Data link layer

The data link layer is responsible for data transfer integrity on an individual link, and ITU-T protocols like LAPB and LAPD do a good job in this regard. It is possible for a link layer disconnect signal to cause the loss of data, but many vendor implementations are designed to account for any outstanding traffic on the link before releasing the session between the two users.

Network layer

A user may issue a clear signal to the network (and the OSI Model's network layer). The purpose of this signal is to request that the user be disconnected from the network and from the other user. It is similar to hanging up the telephone in a telephone network. The effect of the clear is to remove both users from the ability to use the network services.

The reset is somewhat similar to a clear signal, except the reset does not remove the user's connection with the network. It allows the user to inform the network that it wishes to continue the connection but perform certain administrative functions to reinitialize the session with the network and the other user. Once the reset is accomplished, the user can continue its session with the network and the other user.

The restart is like the clear; it is a disconnect. However, the restart may act as a clear for more than one user. While there are exceptions to this statement, the most common use of the restart is to allow the network or the user to reestablish connections from all users on one physical interface to the network (for example, all logical channels on one physical link).

The clear, reset, and restart signals at the network layer may be destructive. The term destructive is used to mean that user data may be lost as a result of these operations. The reader should be aware that networks vary on how they handle these three signals. These possible variances may seem confusing if the goal is to use a standard to define specific activities. However, a network can conform to the X Series Recommendations and still choose options on how to handle these operations.

Transport layer

The transport layer can recover from the potentially destructive signals of clear, reset, and restart. For example, a network level reset might require the transport level to maintain copies of the user data, and after the network reset is complete (with the possible loss of data in the network), the transport layer could perform retransmission functions to recover the data. Therefore, the transport layer is responsible for data transfer integrity across a network or networks.

Notwithstanding this recovery mechanism, it must be emphasized that the transport layer may also perform destructive operations. This event occurs when a user requests a disconnection of its service at the transport layer. The transport layer might then request the network layer for the release. The transport layer does indeed recover from lower layer destructive operations, but it does not recover from its own disconnect. In other words, a disconnect at the transport layer may also be destructive.

Session layer

One of the most important functions of the session layer is the provision for a "graceful close" between the user applications. This service ensures that all data are received before a connection is released at any of the lower layers. The session layer either (1) recovers from a lower layer destructive operation or (2) prevents the transport layer disconnect from being destructive. Therefore, this layer provides for end-to-end integrity of the transfer of data between different computers.

If the user applications reside in the same machine, the session layer can be invoked to provide for data transfer and disconnection integrity. Indeed, if the two applications reside in the same computer, it generally is not necessary to invoke the operations of the lower four layers.

Presentation layer

The issue of data integrity is not applicable to the presentation layer. It has no concern with sequencing, flow control, or acknowledgments and relies on the session layer for a graceful close of a connection.

Application layer

The application layer can be configured with several data integrity functions. One of the standards residing in the application layer, RTSE, can be used to provide for acknowledgments of traffic. Notwithstanding, even this entity relies on the session layer to effect a graceful close. Furthermore, the ACSE participates in the creation of an application context, which is connection-oriented.

Examples of Layer Operations

This section describes how the layers of the OSI Model interact with each other in the same machine and between two machines. Figures 6.15 through 6.19 are used to illustrate these interactions. The purpose of these operations is to transfer data from the originating applications process (AP) to the responding AP. The events are numbered in these figures to assist the reader in following the operations. The numbers are used to identify each specific operation. Table 6.6 is provided to explain the notations in the figures.

These figures show two sublayers at the data link layer: DLC-LLC and DLC-MAC. This implementation is typical of local area network data link layers. For wide area data link layers, the DLC-MAC sublayer is usually not present.

Also, these examples show the use of connection-oriented operations at the network and data link layers with LLC type 2 and X.25, respectively. Connectionless operations are also quite common at these layers.

To begin this discussion, please refer to Figure 6.15. An originating AP will start the process by sending information to its user element (UE). How this information is sent from the AP is not defined by the OSI Model.

Event 1. The UE sends an RT-OPEN request primitive to the RTSE. The notation of (U) indicates that some type of user data is conveyed in this primitive; typically it is obtained from the originating AP. The contents of this data are not used by RTSE; rather, they are intended for use by the responding AP.

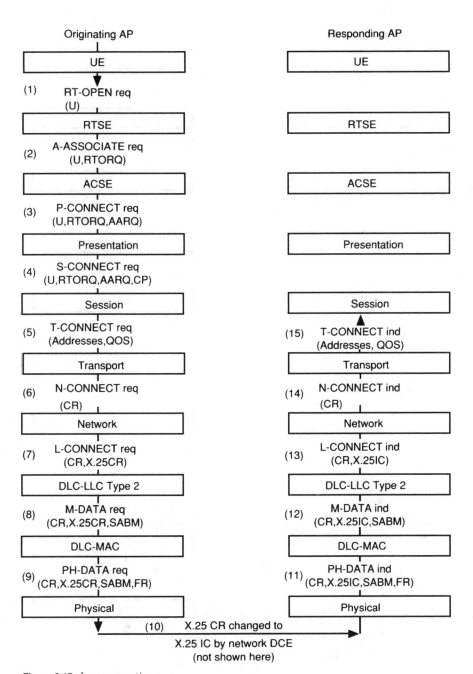

Figure 6.15 Layer operations.

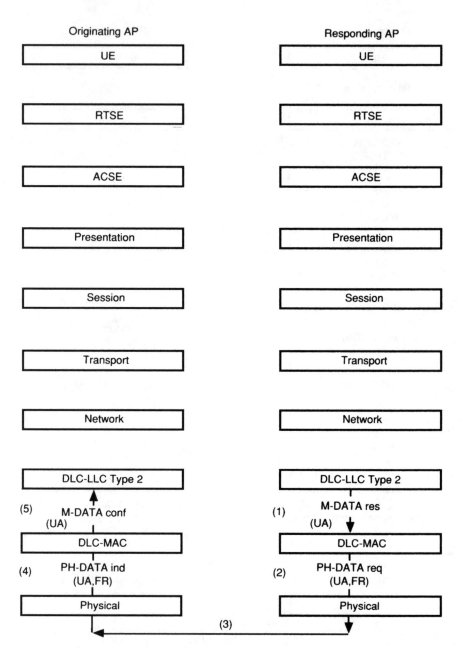

Figure 6.16 Layer operations (*continued*).

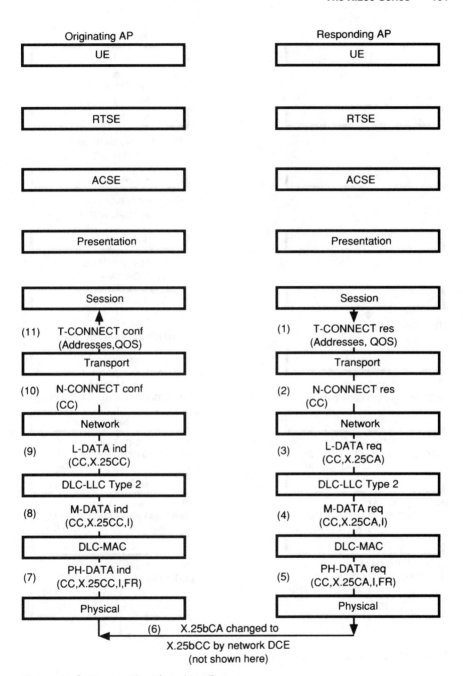

Figure 6.17 Layer operations (*continued*).

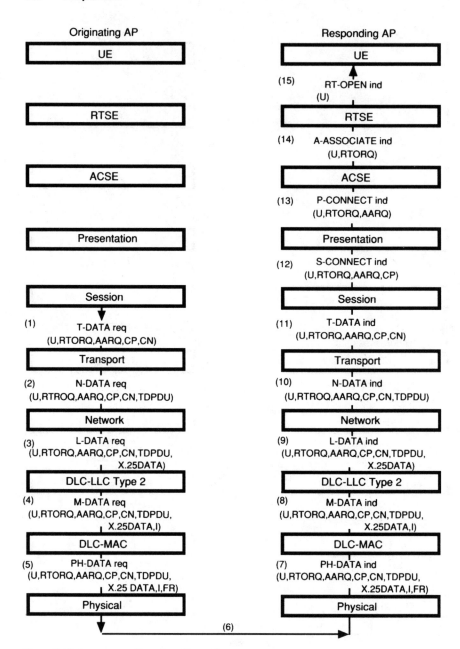

Figure 6.18 Layer operations (*continued*).

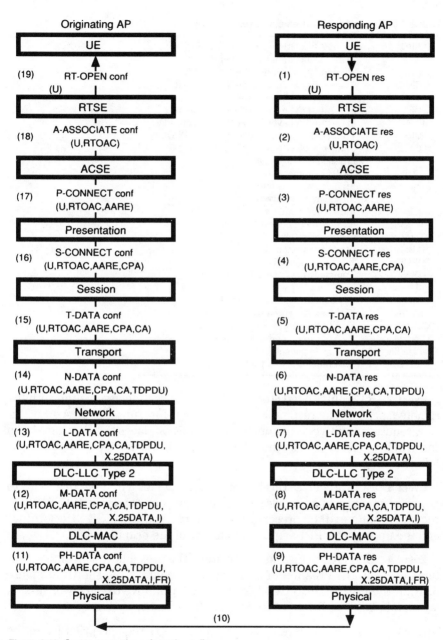

Figure 6.19 Layer operations (*continued*).

TABLE 6.6 Legend for Figures 6.15 through 6.19

AARQ	ACSE Association Open Request PDU
AARE	ACSE Association Open Confirm PDU
CA	Session Layer Connection Accept PDU
CN	Session Layer Connection Request PDU
conf	A confirm primitive
CP	Presentation Layer Connection Request PDU
CPA	Presentation Layer Connection Accept PDU
CR	Transport Layer Connection Request PDU
CT	Connectionless Network Layer Data (DT) PDU
FR	Physical Level Frame
I	SABM I Frame
ind	An indication primitive
req	A request primitive
res	A response primitive
RTORQ	RTSE Open Request PDU
RTOAC	RTSE Open Confirm
SABM	Set Asynchronous Balanced Mode
TDPDU	Transport data PDU
TQOS	Transport Layer QOS Parameters
U	User data
UI	Unnumbered information field
X.25 CA	X.25 Call Accepted packet
X.25 CC	X.25 Call Confirm packet
X.25 CR	X.25 Connection Request packet
X.25 DATA	X.25 Data packet
X.25 IC	X.25 Incoming Call packet

Event 2. RTSE first needs to establish an application association with the remote machine (the responding RTSE). Therefore, it issues an A-ASSOCI-ATE request primitive to the underlying ACSE. It places the U parameter in the primitive call as well as the RTSE PDU, which is labeled in the figure as RTORQ.

Event 3. ACSE receives this information and takes a number of actions to begin the creation of an application context. It then sends the P-CONNECT request primitive to the presentation layer, placing the ACSE PDU (AARQ) and the upper layer PDUs in the parameter of the primitive.

Event 4. Upon receiving this information, the presentation layer takes actions to begin the creation of a presentation context (as shown in Figure 6.10). It then issues the S-CONNECT request primitive to the lower session layer and places its PDU (labeled CP) in the primitive parameters.

Event 5. The session layer issues a T-CONNECT request primitive to the transport layer. Notice that the parameters conveyed down to the transport layer consist of addresses and QOS parameters. The reader should be aware that in actual use how event 5 is implemented can vary. Since the upper layers do not have (as yet) transport and network connections, the T-CONNECT request primitive must convey enough information to the transport layer to enable a connection to be made to the remote machine, as well as to effect a network connection at the next lower level. At a minimum, sending and receiving addresses are needed, as are the QOS features, although the addresses could be derived by vendor-specific means at the transport or lower layers.

Event 6. Most articles and books that discuss this interface simply assume that the transport layer continues the events discussed in the previous layers by creating its own PDU and sending it to the network layer. In this example, it does create a connection request (CR) data unit. Later discussions shows some variations at this interface. Furthermore, it may be necessary to divide the CR PDU into smaller pieces for transport over the network.

The transport layer connection request PDU may be quite lengthy and many packet networks do not support this amount of data in a connection-setup packet at the network layer. From this standpoint, the X.25 fast select feature is quite attractive because it allows the transport layer to send up to 128 octets of data to the network layer. This example does not show the fast select; rather it assumes that a conventional X.25 Call Request connection is to be executed to obtain a network connection. Of course, it is possible that a network connection may already exist (not shown in this example).

The transport layer issues an N-CONNECT request primitive to the network layer. As the figure shows, the transport layer CR PDU is conveyed to the network layer. The QOS parameters are allowed to originate in the upper UE. Since they are concerned specifically with network communications, they are passed transparently down to the transport layer.

The QOS parameters were discussed earlier in this chapter, but for a brief review, they consist of requests such as throughput, transit delay, error rate, etc.

Also, the transport layer could have issued an N-DATA request primitive (instead of the N-CONNECT request) if connectionless service was desired.

Event 7. The network layer passes an L-CONNECT request primitive to the data link layer. The parameter in this packet consists of an X.25 Call Request packet (X.25CR). This packet contains the addresses and QOS parameters passed to the network layer from the transport layer.

As with event 6, the network layer could have issued an L-DATA request primitive (instead of the L-CONNECT request) if connectionless service was wanted at the data link level.

Event 8. The data link layer is using logical link control (LLC) type 2. This protocol is quite similar to LAPB and LAPD, and it is used at this layer to give the reader a fuller explanation of the layering of the lower layers. LLC type 2 is a connection-oriented link layer protocol; therefore, when the LLC passes the M-DATA request primitive to the media access control (MAC) sublayer, the parameters contained in this call consist of the X.25

Call Request packet and the set asynchronous balance mode (SABM) frame. The purpose of the SABM is to establish a connection-oriented link layer session between the originating and responding machines, although a network may act as the conduit between these two. In most operations, the link is set up and remains available for the upper layer operations. If this is the case, the SABM operations would be performed before the operations in this example take place.

Event 9. MAC issues a PH-DATA request primitive and passes the necessary information used to create a frame (FR). The frame is sent across the communications channel to an intervening network and eventually to the responding application process. The FR could be an Ethernet or token ring frame. If so, the SABM PDU need not contain flags nor the frame check sequence (FCS) field. These fields are handled by MAC.

Event 10. The X.25 Call Request packet is changed for the responding machine to an X.25 Incoming Call packet by the network data circuit-terminating equipment (DCE). The contents of the PDU on the communications link as shown in event 10 actually consist of the X.25 Call Request packet encapsulated into the LLC SABM frame, which is in turn encapsulated into the physical link frame of Ethernet, token ring, etc.

Event 11. The data are passed from the physical layer to the receiving MAC through the PH-DATA indication primitive. Notice that the parameters in this primitive are the X.25 Incoming Call (IC), the LLC SABM command, and the frame (FR).

Event 12. MAC passes the M-DATA indication primitive to LLC with the X.25 packet and the LLC SABM command.

Event 13. LLC passes to the network layer the L-CONNECT indication primitive and the X.25 Incoming Call packet. Upon receiving this packet, the responding network layer maps this call into the responding machine.

Event 14. The network layer takes note of the quality of the addresses and service parameters and passes the CR PDU up to the transport layer. As with the sending side, the implementation at this interface varies. The common practice is to pass this primitive to the transport layer along with some of the QOS parameters. The transport layer may not act upon these parameters; in other implementations, it will take some actions. The actual implementation depends upon the vendor's approach in dealing with the QOS parameters between the network and the transport layer, and the ISO and ITU-T consider it a local matter. This interface should be checked carefully when are reviewing vendor's products.

Figure 6.16 shows the activities at the lower layers in establishing the link. Again we emphasize that we are using a SABM-LLC type 2 at the data link layer. Since it is connection-oriented, it must return an unnumbered acknowledgment (UA) frame back to the transmitting machine to verify the establishment of the link. It performs this operation in event 1. The M-DATA response primitive contains the UA frame.

Event 2. MAC passes a PH-DATA request primitive to the physical layer with the UA PDU and the frame.

Event 3. The frame is relayed to the originating machine with the LLC type 2 UA encapsulated into the I field of the frame.

Event 4. The physical layer passes the UA and FR to the MAC layer with the PH-DATA indication primitive.

Event 5. MAC passes the UA PDU to LLC with the M-DATA confirm primitive.

The reader will notice that the operations in events 2 and 4, between MAC and the physical layer, are not using any response and confirm primitives. Therefore, this interface is acting in a connectionless mode. In contrast, in events 1 and 5, the response and confirm primitives, effectively provide a connection-oriented protocol at this layer. Events of 1 and 5 can be changed to be connectionless as well, which is quite common in local area networks.

As a brief summary, the operations in Figure 6.16 have resulted in the establishment of an ongoing communications link between the two machines (or between the machine and intervening network).

The operations in the figure may take place before the operations in Figure 6.15 take place. In many instances, a link is "nailed up" and left up indefinitely as long as it operates properly. Therefore, it serves as a "pipeline" between the machines. Moreover, the network layer connection may also be established before the upper layers are invoked. So, the activities in Figures 6.15 and 6.16 vary, according to specific implementations.

Figure 6.17 shows the operations at the lower layers to complete a transport and network connection. This illustration presumes a one-to-one relationship between the transport and the network connections. This relationship may or may not be the case, so again the reader should check with the vendor to determine how these relationships occur.

Event 1. The session layer sends the T-CONNECT request primitive to the transport layer.

Event 2. The transport layer sends to the network layer the N-CONNECT response primitive with the transport layer CC PDU. If the QOS parameters presented to the transport layer in Figure 6.15 are relevant to the transport layer, the transport layer is allowed to alter them before passing them to the network layer. Generally, the QOS parameters may only be negotiated down; they may not be negotiated up unless they are already below the default level in the standard.

Event 3. The network layer receives the primitive and conveys to LLC an X.25 Call Accepted (CA) packet in the L-DATA request primitive. The packet contains the CC PDU as well.

Event 4. This information is conveyed down to MAC with the M-DATA request primitive. Notice that the CC PDU and the X.25 Call Accepted packet are also carried with the LLC information field (I).

Event 5. MAC sends this information down to the physical layer with a PH-DATA request primitive.

Event 6. The information is transported to the originating machine. Notice that the note in event 5 establishes that the Call Accepted packet is changed to the Call Confirm packet by the network DCE.

Events 7, 8, and 9. Note that the information is transported back up the layers with the appropriate indication (ind) primitives.

Event 10. The network layer sends to the transport layer the CC PDU in the N-CONNECT confirm primitive.

Event 11. The T-CONNECT confirm primitive is conveyed to the session layer.

Next, Figure 6.18 is examined. It can be seen that the operations at the upper layers of the originating AP in Figure 6.14 are finally used at the upper layers of the responding application layer.

Event 1. The session layer sends a T-DATA request primitive to the transport layer. The session layer has been holding the U, RTORQ, AARQ, CP, and CN PDUs. It now sends them to the transport layer.

Event 2. The transport layer relays these PDUs to the network layer through the N-DATA request primitive. It places them in the transport data protocol data unit (TDPDU).

Event 3. The network layer conveys this information to LLC through the X.25 data packet (X.25 DATA).

Events 4 through 10. This information is sent down to the lower layers of the originating machine and up through the lower layers of the responding machine.

Event 11. The transport layer passes the upper layer protocol parameters to the session layer through the T-DATA indication primitive.

Event 12. The session layer examines the CN PDU to determine what actions the peer session layer in the originating machine wishes to negotiate. It then passes the remainder of the parameters up the presentation layer through the S-CONNECT indication primitive, which contains the U, RTORQ, AARQ, and CP data units.

Event 13. The presentation layer examines its CP PDU, performs actions as defined in the CP header, and passes the remaining parameters to the next layer.

Event 14. The ACSE examines the AARQ header, takes appropriate actions, and passes the remainder of the parameters to the next layer. This operation take place with the A-ASSOCIATE indication primitive.

Event 15. RTSE examines the RTORQ and passes the last parameter up to UE through the RT-OPEN indication primitive.

Finally, Figure 6.19 shows the procedures involved in the responding machine sending to the originating machine the required DUs to complete the end-to-end association. The reader should be able to follow this figure with ease, since it is based on the previous four examples and figures. The only noteworthy aspects of this figure that differ from the other examples show that the responding RTSE, ACSE, and presentation and session layers return confirm PDU identified as RTORC, AARE, CPA, and CA, respectively, in the figure. Table 6.6 explains what these initials mean.

After the activities shown in Figure 6.19 are complete, the APs can begin to transfer data to each other. How the data are exchanged depends upon how the AP session was negotiated at the seven layers during the activities that took place in this example.

Summary

The ITU-T X.200 Recommendations represent some of the more lengthy documents in the Blue Books. These standards define the operations for data networks using the OSI Model. The operations between each layer are defined with service definitions; the operations between peer layers in different machines are defined with protocol specifications. The ITU-T now publishes protocol service definitions for each of the seven layers of the OSI Model.

The lower layers of this model are relatively mature. The upper layers, and especially the application layer, are still undergoing rather dramatic and frequent changes.

Note: The absence of protocols at a layer does not mean the layer contains no protocols. Rather, it means other ITU-T Recommendations are used to fulfill the functions of the layer, such as other X Series, or protocols in the V, I, and Q Series.

7

The X.300 Series

Data transmission services are now offered by different types of networks. These networks often must communicate with each other in order to provide an end-to-end service to a user. To provide a common framework for internetwork operations, the ITU-T publishes a series of standards pertaining to the internetworking of various types of networks. They encompass several documents numbered X.300 through X.370, generally called the X.300 Interworking Recommendations. You may have noticed that we have used the word, interworking. The ITU-T uses this word in place of *internetworking*. This chapter examines the most important aspects of the X.300 Recommendations. The organization of the X.300 Series is shown in Figure 7.1, and the full titles are listed below.

- X.300: General principles for interworking between public networks, and between public networks and other networks for the provision of data transmission services

- X.301: Description of the general arrangements for call control within a subnetwork and between subnetworks for the provision of data transmission services

- X.302: Description of the general arrangements for internal network utilities within a subnetwork and intermediate utilities between subnetworks for the provision of data transmission services

- X.305: Functionalities of subnetworks relating to the support of the OSI connection-mode network service

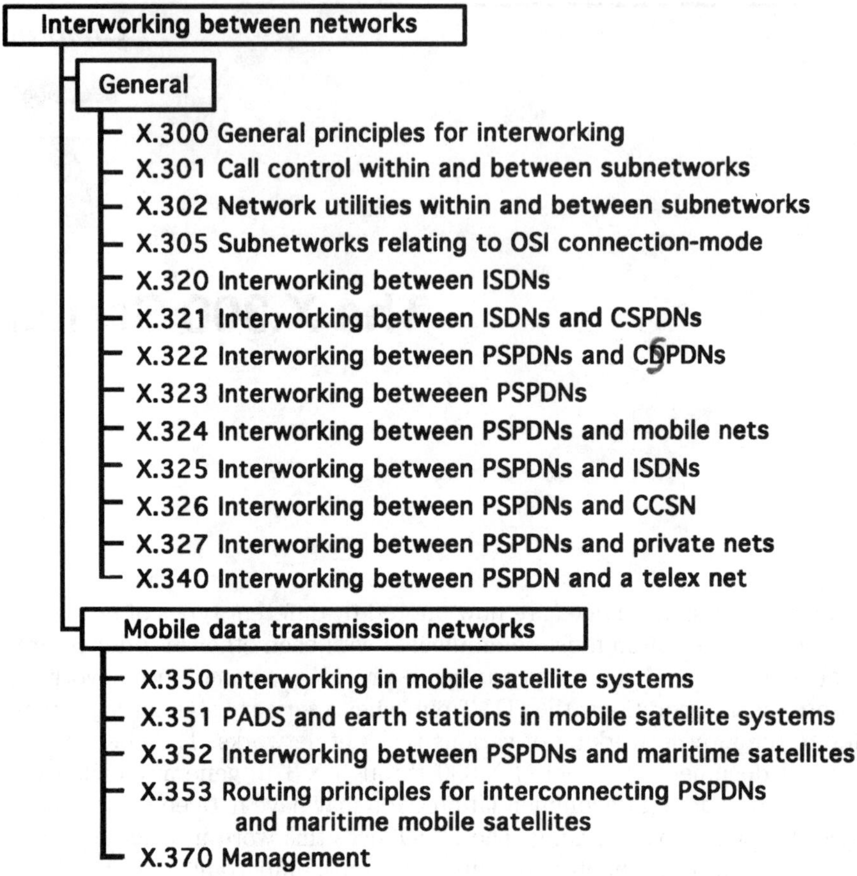

```
┌─────────────────────────────────────────────┐
│ Interworking between networks                 │
└─────────────────────────────────────────────┘
    ┌──────────────┐
  ──│ General       │
    └──────────────┘
     ├─ X.300 General principles for interworking
     ├─ X.301 Call control within and between subnetworks
     ├─ X.302 Network utilities within and between subnetworks
     ├─ X.305 Subnetworks relating to OSI connection-mode
     ├─ X.320 Interworking between ISDNs
     ├─ X.321 Interworking between ISDNs and CSPDNs
     ├─ X.322 Interworking between PSPDNs and CSPDNs
     ├─ X.323 Interworking betweeen PSPDNs
     ├─ X.324 Interworking between PSPDNs and mobile nets
     ├─ X.325 Interworking between PSPDNs and ISDNs
     ├─ X.326 Interworking between PSPDNs and CCSN
     ├─ X.327 Interworking between PSPDNs and private nets
     └─ X.340 Interworking between PSPDN and a telex net
    ┌─────────────────────────────────────┐
  ──│ Mobile data transmission networks     │
    └─────────────────────────────────────┘
     ├─ X.350 Interworking in mobile satellite systems
     ├─ X.351 PADS and earth stations in mobile satellite systems
     ├─ X.352 Interworking between PSPDNs and maritime satellites
     ├─ X.353 Routing principles for interconnecting PSPDNs
     │         and maritime mobile satellites
     └─ X.370 Management
```

Figure 7.1 Organization of the X.300 Series.

- X.320: General arrangements for interworking between integrated services digital networks (ISDNs) for the provision of data transmission services

- X.321: General arrangements for interworking between circuit switched public data networks (CSPDNs) and integrated service digital networks (ISDNs) for the provision of data transmission services

- X.322: General arrangements for interworking between packet switched public data networks (PSPDNs) and circuit switched public data networks (CSPDNs) for the provision of data transmission services

- X.323: General arrangements for interworking between packet switched public data networks (PSPDNs)

- X.324: General arrangements for interworking between packet switched public data networks (PSPDNs) and public mobile systems for the provision of data transmission services

- X.325: General arrangements for interworking between packet switched public data networks (PSPDNs) and integrated services digital networks (ISDNs) for the provision of data transmission services

- X.326: General arrangements for interworking between packet switched public data networks (PSPDNs) and common channel signalling network (CCSN)

- X.327: General arrangements for interworking between packet switched public data networks (PSPDNs) and private data networks for the provision of data transmission services

- X.340: General arrangements for interworking between a packet switched public data network (PSPDN) and the international telex network

- X.350: General interworking requirements to be met for data transmission in international public mobile satellite systems

- X.351: Special requirements to be met for packet assembly/?disassembly facilities (PADs) located at or in association with coast earth stations in the public mobile satellite service

- X.352: Interworking between packet switched public data networks and public maritime mobile satellite data transmission systems

- X.353: Routing principles for interconnecting public maritime mobile satellite data transmission systems with public data networks

- X.370: Arrangements for the transfer of internetwork management information

The reader should review the abbreviations in Table 7.1 before reading this material. It will also prove beneficial to review the X.213 protocol specification in chapter 6.

TABLE 7.1   Abbreviations Used in X.300

CCSN	Common channel signalling network (SS No. 7)
CSPDN	Circuit switched public data network
IDSE	International data switching exchange
IWF	Interworking function
MSS	Maritime satellite service
NDSE	National data switching exchange
PDN	Public data network
PSPDN	Packet switched public data network
PSTN	Public switched telephone network
TDI	Transit delay indication
TDS	Transit delay selection
TDSAI	Transit delay selection and indication
TOA	Type of address

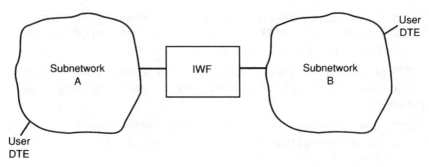

Figure 7.2 The interworking function.

The Interworking Function (IWF)

The ITU-T uses the term *interworking function* to describe what most people in the industry call a gateway. This device typically operates at the lower three layers of the OSI Model (in contrast to a bridge, which operates at the lower two layers). The IWF performs routing functions (and perhaps quality of service operations) between networks. Figure 7.2 shows an IWF placed between network A and network B.

Networks A and B are called subnetworks. The term does not mean that they provide fewer functions than a conventional network. Rather, it means that the two networks consist of a full logical network with the subnetworks contributing to the overall operations for interworking. Stated another way, the subnetworks comprise an internet, although the ITU-T does not employ this widely used term.

An IWF is designed to remain transparent to the end-user application. Indeed, the end-user application resides in the host machines connected to the networks and operates above the application layer of the OSI Model.

In addition to application layer transparency, most designers attempt to keep the gateway transparent to the subnetworks and vice versa. That is to say, the gateway does not care what type of network is attached to it. The principal need is for the gateway to receive a protocol data unit (PDU) that contains adequate addressing information to enable the gateway to route the PDU to the proper final destination or the proper next node gateway.

X.300 and the Other X Series Recommendations

Perhaps one of the most useful features of X.300 is its explanation of how the X Series Recommendations fit into interworking. Figure 7.3 is a view of the X.300 X Series framework. As this figure suggests, the standards are divided into three categories:

- Aspects of interworking pertinent to different cases
- Aspects of interworking pertinent to each case
- Aspects of interworking signalling interfaces

This book examined the signalling interface standards in previous sections and chapters. Our task here is to examine the X.300 specifications.

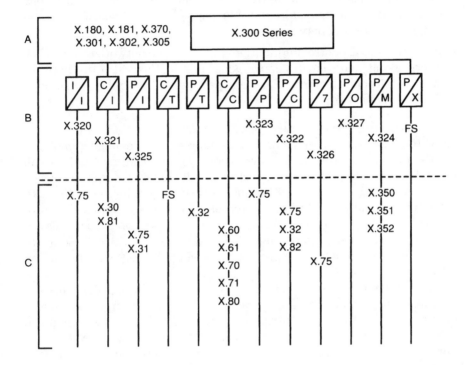

Legend:
A: Arrangements generic to different cases
B: Arrangements for each case
C: Interwork signalling interfaces

Abbreviations:
C: Circuit switched data network
I: ISDN
M: Mobile network
O: Privately owned network
P: Packet switched data network
T: Telephone network
X: Telex network
7: Signalling system no. 7
FS: For further study

Figure 7.3 The X.300 framework.

Categories of Interworking

The X.300 transmission capability between two networks is divided into two categories. One category uses a convergence protocol, which rests on top of a subnetwork (transparent to that network) in order to support yet another network service. The categories are:

- Call control mapping: Call control information used for switching in one subnetwork (including addressing) is mapped into call control information in the other subnetwork.

- Port access: Call control information is used by one network to select the interworking point. Then, a convergence protocol is used at this network and the call control information is mapped into call control information in the other subnetwork.

In addition to these two categories, X.300 also defines subnetworks in relation to Recommendation X.213. Several tables in Recommendation X.300 are used to (1) identify subnetwork types, (2) establish the need for call control mapping or port access, (3) stipulate the requirement for a convergence protocol, (4) show resulting subnetworks from interconnecting two subnetworks, and (5) define the different subnetwork types used to provide an OSI connection-mode network service (connectionless services are not defined in X.300). Interestingly, ITU-T does not provide a descriptive definition of the subnetwork types but relies on the X.300 Recommendation tables to define them.

This information is provided as shown in Tables 7.2 through 7.4. Table 7.2 identifies four network types based on the X.213 connection establishment, data transfer, and connection release phases. Table 7.3 shows when the subnetwork interconnecting requires call control mapping and when it requires port access. Finally, Table 7.4 stipulates whether or not a convergence protocol is required by the four network types for the X.213 phases.

TABLE 7.2 Subnetwork Types

Subnetwork type	Phase of the call		
	Connection establishment	Data transfer phase	Connection release phase
I	M	M	M
II	M	P	M
III	S	P	S
IV	M or S	F	M or S

M = all mandatory elements required for the provision of the OSI network service are signalled through the subnetwork by means of its signalling capability.

P = the functionality of the subnetwork corresponds to that of a physical connection.

TABLE 7.2 (Continued)

F = some form of packetizing or framing is provided by the subnetwork, but all mandatory services needed for an OSI network service may not be provided.

S = a subset of all mandatory elements required for the provision of the OSI network service are signalled through the subnetwork by means of its signalling capability.

TABLE 7.3 Interconnecting Two Networks: Categories of Internetworking

Subnetwork type	Subnetwork type			
	I	II	III	IV
I	Internetworking by call control mapping	Internetworking by call control mapping or by port access	Internetworking by call control mapping or by port access	Internetworking by call control mapping or by port access
II	Internetworking by call control mapping or port access	Internetworking by call control mapping	Internetworking by call control mapping or by port access	Internetworking by call control mapping or by port access
III	Internetworking by call control mapping or by port access	Internetworking by call control mapping or by port access	Internetworking by call control mapping	Internetworking by call control mapping or by port access
IV	Internetworking by call control mapping or by port access	Internetworking by call control mapping or by port access	Internetworking by call control mapping or by port access	Internetworking by call control mapping

TABLE 7.4 Providing OSI Connection-Mode Network Service with Different Subnetwork Types

Subnetwork type	Phase of the OSI-NS call connection		
	Connection establishment phase	Data transfer phase	Connection release phase
I	No Convergence Protocol Required	No Convergence Protocol Required	No Convergence Protocol Required
II	No Convergence Protocol Required	Convergence Protocol Required	No Convergence Protocol Required
III	Convergence Protocol Required	Convergence Protocol Required	Convergence Protocol Required
IV	Convergence Protocol Required*	Convergence Protocol Required	Convergence Protocol Required

*If this subnetwork does not provide all the mandatory elements of the OSI network service in this phase.

The ITU-T categorizes the following networks as type I networks: PSPDN, integrated services digital network (ISDN) packet-switched-based, and mo-

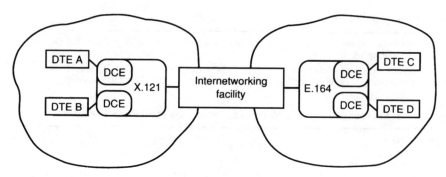

Figure 7.4   X.301 functional domain.

bile satellite systems (MSS). An ISDN-circuit-switched-based network is an example of a type II network. Type III networks are CSPDNs and PSTNs. Presently, a type IV network does not exist under the ITU-T scheme.

Transfer of Addressing Information

One of the more important parts of the X.300 and X.370 Recommendations is contained in X.301: rules for the transfer of addressing information. We learned about the various addressing schemes in chapter 5. X.301 defines how X.121 and E.164 addresses are used between two networks.

Figure 7.4 shows the X.301 structure for the address form when the IWF is interworking an X.121 address-based network and an E.164 address-based network. Table 7.5 shows the permissible address forms for the call establishment phase. The prefixes P1 through P6 are not passed over the IWF, and their form and use are an internal network matter. The escape digits E1 and E2 indicate that the succeeding address is a different numbering plan. The prefixes may or may not precede the escape digit.

It is also permissible to use another field to indicate the type of address present. This element in the protocol is called the *Number Plan Indicator /Type of Address (NPI/TOA)*. The form of this field depends on the specific network access protocol employed.

Cases for Interworking with Specific Systems

The X.300 standards that pertain to specific systems ("for each case") are described in this section. X.25-based packet network systems represent most of the parts of these X.300 documents. They are as follows:

- X.322: The case of circuit switched public data networks (CSPDNs)
- X.323: The case of packet switched public data networks (PSPDNs)

- X.324: The case of mobile systems
- X.325: The case of integrated services digital networks (ISDNs)
- X.326: The case of common channel signalling networks (CCSNs)
- X.327: The case of private data networks
- X.340: The case of telex networks

The recommendations that do not pertain to X.25 are:

- X.320: The case of interworking ISDNs
- X.321: The case of interworking circuit switched public data networks with ISDNs

The X.300 Recommendations do not provide a specific document for the following interworking cases:

- The case of interworking circuit switched and telephone networks
- The case of interworking packet switched and telephone networks
- The case of interworking circuit switched networks

X.324, X.350 and X.353: Interworking X.25 and Maritime Satellite Systems

In the past, ships at sea were equipped with high-frequency (HF) radio

TABLE 7.5 Call Establishment Phase Address Forms

Direction	Form of address	Extent of validity
A to B	NTN	Network
A to B	P1 + NTN	Network
A to B	DNIC + NTN	Internetwork
A to B	P2 + DNIC + NTN	Internetwork
A to B	NTN + [NPI/TOA]	Network
A to B	DNIC + NTN + [NPI/TOA]	Internetwork
C to D	SN	Network
C to D	P3 + SN	Network
C to D	CC + (NDC) + SN	Internetwork
C to D	P4 + CC +(NDC) + SN	Internetwork
C to D	SN + [NPI/TOA]	Network
C to D	CC + (NDC) + SN + [NPI/TOA]	Internetwork
A to C	E1 + CC + (NDC) + SN	Internetwork escape to E.164/E.163
A to C	P5 + E1 + CC + (NDC) + SN	Internetwork escape to E.164/E.163
A to C	CC+ (NDC) + SN + [NPI/TOA]	Internetwork
C to A	E2 + DNIC + NTN	Internetwork escape to X.121
C to A	P6 + E2 + DNIC + NTN	Internetwork escape to X.121
C to A	DNIC + NTN + [NPI/TOA]	Internetwork

systems, using telephony and telex transmission schemes. The earlier systems also relied on Morse code. While HF systems are still used (with signal reflection across the ionized layers), the majority of maritime users are using satellite systems for transmission and reception of data, video, and voice.

In 1979, the International Maritime Satellite Organization (INMARSAT) was founded to foster standards and operational facilities for at-sea communications. It became operational in 1982 and now provides continuous communications services to equipped ships in all parts of the world.

The operations of INMARSAT are governed by several ITU-T Recommendations. In this section we discuss the operations of a maritime-ship communication system as well as the ITU-T mobile data transmissions specifications.

An at-sea ship's communications process with a satellite is controlled by an earth station (obviously a misnomer in this case). The ship's antenna dish is quite small, only 0.9 meter in diameter. It is mounted on the superstructure of the ship to prevent interference with the signal. In order to maintain alignment with the satellite, a stabilization system is used to keep the dish pointed directly at the communicating satellite.

X.324

The X.324 recommendation is a very brief document that describes the relationships of these specifications and defines the requirements for the use of other standards, such as X.1, X.2, and X.10 (see chapter 3 for a description of these standards).

X.350

This recommendation defines the maritime satellite circuits and the addressing rules for international data numbers. Figure 7.5 shows the topology for these circuits.

The maritime local circuit is between the ship's DTE and the ship's satellite station (MES for mobile earth station). The maritime satellite circuit is the satellite channel between the ship's earth station and the coastal earth station (CES). The maritime terrestrial circuit is the circuit connecting the CES to the maritime satellite data switching exchange (MSDSE). This exchange provides the internetworking between the satellite system and the public data network. It handles routing as well as call control to and from ships at sea. It is also responsible for managing the charging of the calls.

X.350 specifies several options for the interface between the components. For example, Recommendations X.21, X.21 bis, and X.22 are permitted for circuit-switched public data networks and X.25 is permitted for interfacing into packet networks. ITU-T recommends that the interface be provided through an X.25 public data network because of the multiplicity of

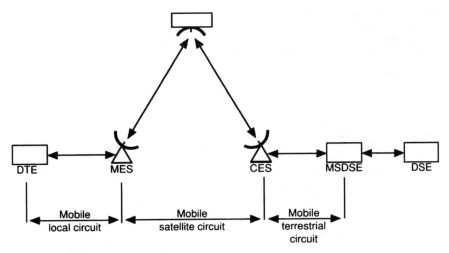

Figure 7.5 Maritime satellite topology.

functions provided in this specification and its worldwide use.

Recommendation X.121 is used to address the ships at sea. The format for the shipboard data terminal equipment (DTE) is composed of: the data network identification code (DNIC) plus a ship station identity (Recommendation E.210-F.120) plus optional digits identifying the on-board DTE. Three DNICs have been assigned the Atlantic, Pacific, and Indian Oceans. These numbers are 1111, 1112, and 1113, respectively.

In addition to the X.121 identification scheme, the actual types of transmissions are identified by prefix codes. Presently, a two-digit code is used to describe the types of calls, such as telephone, telex, or data, and the category of calls within those transmission types. Some examples for the use of prefix codes are shown in Table 7.6.

X.351

The recommendation defines the procedures for the use of packet assembly/disassembly operations that operate with the CES. Generally, the operations are in conformance with X.3, X.28, and X.29 (see chapters 3 and 4 for a discussion of these recommendations). The parameters in the X.3 table are defined in this recommendation for maritime PADs.

X.352

This recommendation describes the layers and the protocols involved in the mobile satellite call setup and disconnection. As the reader might surmise, X.21, X.21 bis, and the V Series are cited for use at the physical layer. Link

TABLE 7.6 Examples of Maritime Satellite Service, Prefix and Access Codes

Code	Application
15	Radiogram service
20	Access-to-maritime PAD
24	Telex letter service
32	Medical advice
34	Person-to-person call
36	Credit card calls
38	Medical advice
39	Maritime assistance
41	Meteorological reports
42	Navigational hazards and warnings
43	Ship position reports
51	Meteorological forecasts
55	News, international
56	News, national

access protocol, balanced (LAPB) is cited for use at the data link layer, with an option to use the selective reject (SREJ) command frame to request for retransmissions. X.25 packet layer procedures are used at the network layer. Annex A of X.352 provides some informative and useful illustrations of all the signals involved in the call.

X.353

This recommendation provides some rather general and terse descriptions of routing principles for maritime mobile satellite systems. X.353 establishes the following rules:

- The MSDSE acts as a gateway for the MESs.
- The MSDSE may serve more than one satellite and may serve more than one mobile system, as well as multiple packet networks.
- Routing is based on the use of the X.121 DNICs.

The title of this recommendation is somewhat inappropriate in that it does not provide much information about how to route the traffic. It contains no routing guidelines and certainly no routing flow charts or algorithms.

X.323 Interworking Packet Networks

This specification is also quite terse. It simply sets the requirements for the use of X.75 for the interworking arrangements between two packet networks. As with the other X.300 specifications, it describes the use of

user classes of service with X.1 and user facilities with X.2. It also stipulates the use of X.29, X.31, X.32, and X.96 for this internetworking
arrangement. In addition, it stipulates the use of logical links A1 and G1
as defined in X.92.

X.325 Interworking ISDNs and Packet Networks

This specification establishes the procedures for internetworking ISDNs
and packet networks that use the X.25 interface. X.325 requires the use of
X.72 and X.31 for the specific internetworking signalling interfaces. As the
reader might expect at this point in the discussion of X.300 Recommendations, X.325 also stipulates the specific transmission services as defined in
X.1 and X.2. X.325 is a bit more detailed than some of its counterparts.

Since ISDNs and packet networks use different addressing and numbering plans (for example E.164 and X.121), X.301 describes the mapping of
the addresses between the two different types of networks. This mapping
was described earlier in this chapter.

X.326 Interworking X.25 with Common
Channel Signalling Networks (CCSN)

X.326 defines the internetworking between CCSNs and packet networks
in the context of using the OSI Model. X.326 recommends that the packet
network offer the full capability of the OSI network layer service (as defined in X.213 and X.223). Furthermore, the standard stipulates that an
OSI connection-oriented network layer service is to be provided for the
interworking of these two networks. X.326 stipulates the use of X.75 as
the internetworking protocol between the packet networks and the common channel signalling networks.

X.327 Interworking X.25 with Private Data Networks

This specification, as might be expected, has little to say about interworking
between public and private networks since the ITU-T's orientation is toward
public internetworking. However, X.327 does provide guidance on the interconnection of X.25 devices through public and private X.25-based networks.
Basically, the major aspect of X.327 is the description of the mapping of connection-oriented services (as defined in X.213) to the X.25 packets.

X.340 Interworking X.25 with Telex Networks (Defunct)

X.340 was removed as an X Series Recommendation with the publication of
the Blue Book. It provided the rules for the use of the interworking unit, as
well as the use of telex-based features. It provided a description of the map-

ping between the X.3 specifications and a telex terminal. The reader can refer to the Red Book if this information is needed.

X.320 Interworking ISDNs

This is a rather terse specification that defines the procedures for ISDN interworking. It establishes the relationships of the interconnection with X.75 or SS7.

X.321 Interworking Circuit-Switched Networks and ISDNs

This recommendation is similar to X.320. It has not been completed as of the drafting of the Blue Book working documents and is subject to "further study."

X.370 Internetwork Management

The recommendation provides a very general and terse description of internetwork management schemes. It references X.213, X.214, and X.224 for more detailed information on the subject. Of course, the Blue Book was published after the release of the X.700 Recommendations (see chapter 11). It is likely that this recommendation will be deleted or modified in the future due to X.700.

Summary

The X.300 Recommendations provide some very useful guidelines on internetworking different kinds of networks. The emphasis in the X.300 series is on packet-switched networks and ISDNs, as well as satellite systems. Perhaps the most useful part of the X.300 series is the definitions on interworking, as well as the various categories and guidelines on mapping addresses between networks. These recommendations are likely to be expanded significantly as more experience is gained about their use.

The X.400 Series

The X.400 Message Handling Systems (MHS) Recommendation began with the initiative of several organizations in North America that had worked with message handling systems in the ARPAnet environment. The initial standards work was assumed by the ITU-T, which published the X.400 Series in 1984. Other organizations have sponsored somewhat complementary standards. For example, the European Computer Manufacturers Association (ECMA) produced the ECMA 93 Distributed Application for Message Exchange (MIDA). It was similar to the 1984 ITU-T version of MHS and offered a few extra services. Organisation International de Normal-isation (International Organization for Standardization, ISO) also came into the picture in 1984. Their initial standards were considerably different from that of ITU-T, but various compromises were made and in the 1988 Blue Books, very close harmonization has been achieved between the ITU-T and the ISO versions.

Introduction

It is impossible to explain MHS in any detail in one short chapter. Therefore, our goal is to describe its major aspects. The reader who wishes only a general overview of the X.400 Recommendations should read the following sections: "The Structure of MHS," "The Architecture of MHS," and "MHS Service Elements." The section titled "The P Protocols" is written for the reader who wishes to know more details about MHS.

```
┌────────────────────────────────────┐
│ Message Handling Systems (MHS)      │
└────────────────────────────────────┘
  ├─ X.400 System and service overview
  ├─ X.402 Overall architecture
  ├─ X.403 Conformance testing
  ├─ X.407 Abstract service definition conventions
  ├─ X.408 Encoded type conversion rules
  ├─ X.411 Message transfer system
  ├─ X.413 Message store
  ├─ X.419 Protocol specifications
  ├─ X.420 Interpersonal messaging system
  ├─ X.435 Electronic data interchange (EDI)
  ├─ X.440 Voice messaging system
  ├─ X.480 MHS and directory services-conformance testing
  ├─ X.481 P2-Protocol implementation conformance statement
  ├─ X.482 P1-Protocol implementation conformance statement
  ├─ X.483 P3-Protocol implementation conformance statement
  ├─ X.484 P7-Protocol implementation conformance statement
  └─ X.485 Voice messaging protocol implementation conformance statement
```

Figure 8.1 Organization of the X.400 Series.

The organization of the X.400 Series is shown in Figure 8.1, and the full titles are listed below.

- X.400: Message handling systems and service overview

- X.402: Message handling systems: Overall architecture

- X.403: Message handling systems: Conformance testing

- X.407: Message handling systems: Abstract service definition conventions

- X.408: Message handling systems: Encoded information type conversion rules

- X.411: Message handling systems: Message transfer system: Abstract service definition and procedures

- X.413: Message handling systems: Message store: Abstract-service definition

- X.419: Message handling systems: Protocol specifications

- X.420: Message handling systems: Interpersonal messaging system

- X.435: Message handling systems: Electronic data interchange messaging system

- X.440: Message handling systems: Voice messaging system

- X.480: Message handling systems and directory services-Conformance testing

- X.481: P2 protocol: Protocol implementation conformance statement (PICS) proforma

- X.482: P1 protocol: Protocol implementation conformance statement (PICS) proforma

- X.483: P3 Protocol: Protocol implementation conformance statement (PICS) proforma

- X.484: P7 Protocol: Protocol implementation conformance statement (PICS) proforma

- X.485: Message handling systems: Voice messaging system protocol implementation conformance statement (PICS) proforma

Because many of the X.400 Recommendations are closely coupled together, this chapter discusses them as a logical whole, and not as individual specifications as in previous chapters. In addition, because X.450-X.485 follow most of the conventions of OSI Conformance testing, these concepts are explained in chapter 13.

The Structure of MHS

The MHS recommendations provide message services with two principal features for end users (originators and recipients of the messages): (1) The *message transfer* system (MTS) supports application-independent systems. It is responsible for the "envelope" of the letter. The MTS is the means by which the users exchange messages. (2) The *interpersonal messaging* (IPM) service supports communications with existing ITU-T, telex, and telematic services and defines specific user interfaces with MHS. It is responsible for the contents within the envelope. It uses the capabilities of the MT service to send and receive messages. MHS considers IPM service to be a standardized application.

MHS uses the term *functional objects* to describe the components of the system. For example, the message handling system users and message distribution lists are classified as primary functional objects. Message transfer systems, user agents, message stores, and access units are classified as secondary functional units.

The Architecture of MHS

The following entities are the major parts (objects) of an ITU-T message handling system (see Figure 8.2):

- User agent
- Message transfer agent

- Message transfer system
- Message store
- Access units

The *user agent* (UA) is responsible for directly interfacing with the end user. MHS does not define how it interacts with the end user or how it performs solitary actions because it is an applications process. In the "real world" it prepares, submits, and receives messages for the user. It also provides text editing and presentation services for the end user. It provides for other activities, such as user-friendly interaction (for example, selective viewing), security, priority provision, delivery notification, and distribution of subsets of documents. Additionally, different types of UAs may be used to provide service to an MHS user. However, as just stated, the interpersonal messaging (IPM) system has been standardized under the X.400 Recommendations, and a unique interpersonal message (called P2) identifies all interpersonal messages.

MHS also defines how the UA interacts with the *message transfer agent* (MTA). UAs are grouped into classes by MHS, based on the types of messages they can process. These UAs are then called cooperating UAs.

The MTA provides the routing and relaying of the message. This function is responsible primarily for the store-and-forward path, channel security,

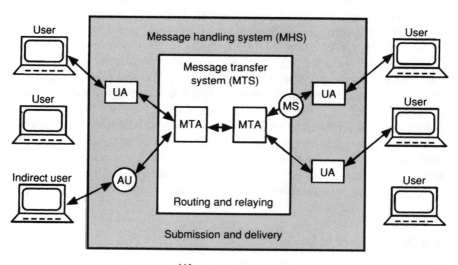

UA = user agent
MTA = message transfer agent
MS = message store
AU = access units

Figure 8.2 X.400 MHS functional view.

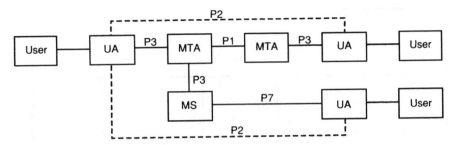

Figure 8.3 The P operations and the MHS functional objects.

and the actual message routing through the communications media. A collection of MTAs is called the *message transfer system* (MTS). These functions are usually specialized to a particular vendor's product, but recent efforts have pointed the way to more standardized systems. The MTS neither examines nor modifies the information part of the message, except for conversion of the *encoded information type* (ASCII code, Group 3 FAX, etc.). Another important point is that MHS considers the MTS to be application independent. Additionally, MTAs can perform application layer routing.

The *message store* (MS) provides for the storage of messages and support their submission and retrieval. The MS complements the UA for machines such as personal computers and terminals that are not continuously available. The job of MS is to provide storage that is continuously available.

Finally, the *access units* (AUs) support connections to other types of communications systems, such as Telematic services, postal services, etc.

The P operations and the MHS objects

The MHS Recommendations contain several protocols that are used for communications between the MHS objects. Figure 8.3 shows the relationships of these protocols to the functional objects:

- P1: Specifies the communications between MTAs
- P3: Specifies the submission and delivery operations between MTAs and UAs
- P7: Specifies the submission and retrieval to and from the MS
- P2: Specifies the interpersonal message protocol between IPM UAs

Contents and the envelope

As shown in Figure 8.4, MHS uses the term *content* to describe the information of the message. It is analogous to a letter. The term *envelope* describes control information used to effect the delivery of the content. It is

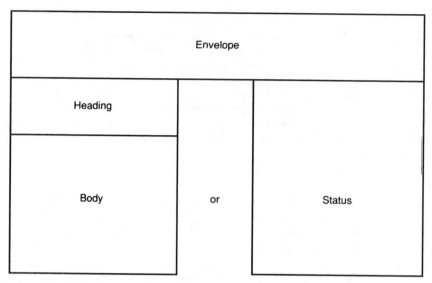

Figure 8.4 The MHS message and envelope.

analogous to the envelope we use in our postal mail service, except an MHS envelope contains the information the MTA needs to invoke many service elements that the postal service does not perform.

MHS also supports the use of *distribution lists* (DLs). A DL is similar to a distribution list in a typical office. For example, an MHS message originator may provide MHS the name of the DL. It is the task of MHS to "expand" the list to the recipient names and distribute a copy of the message to each recipient user.

MHS messages, probes, and reports

The MHS conveys information of three types: *messages, probes,* and *reports.* The structure of the MHS messages consists of (1) an envelope that carries the information needed to transfer the message and (2) the content, which is the information the originating UA wishes delivered to one or more UAs. The contents consist of a heading and a body, which are collectively called the interpersonal message (IP). The MTS neither examines nor modifies the content.

One piece of information carried by the envelope is the type of the content, which is an ASN.1 Object Identifier or Integer type. This information denotes the syntax and semantics of the content. It is used by the MTS to assess the message's deliverability to users.

A probe is an information object that only contains an envelope. It is delivered to the MTAs serving the end users. It is used to determine if the message can be delivered, because it elicits the same behavior from the receiving MTS as would a submitted message. This handy feature is like saying, "I am considering sending you correspondence. These are its characteristics. Will you accept it?" In this way, a lengthy message or a message that requires substantial processing is less likely to be rejected by the receiver.

The report is a status indicator. It is used by the MTS to relate the progress or outcome of a message's or probe's transmission to one or more potential users.

MHS places rules on which functional objects are the ultimate sources or ultimate destination of messages, probes, and reports. These rules are listed in Table 8.1.

The MHS specifications—An overview

X.400 is an involved and detailed specification. An overview of each recommendation is provided in this section.

X.400 describes the basic MHS model in accordance with the OSI Reference Model. X.400 describes in very general terms how an originator interacts with the UA system to prepare, edit, and receive messages. It describes how the UA interacts with the MT network. X.400 describes the interaction between the UA entity and the MT entity and also describes naming and addressing conventions. This document provides a very valuable summary of the all-important MHS *elements of service,* and we will pay considerable attention to this aspect of MHS in a later section.

X.402. Describes the overall architecture of MHS and gives examples of possible physical configurations. This specification contains some very useful definitions and rules for naming and addressing. The interested reader will find it a good reference point for the study of X.400.

X.403. Provides directions on conformance testing. It is a very detailed and lengthy document that defines conformance requirements, testing methodology, test structures, timers, protocol data units, and so forth.

X.407. Specifies conventions used in a distributed information processing task. It describes these tasks in an abstract manner.

TABLE 8.1 Conveyable Information Objects

| Information object | Functional object | | | | |
	User	UA	MS	MTA	AU
Message	S/D	—	—	—	—
Probe	S	—	—	D	—
Report	D	—	—	S	—

S = ultimate source
D = ultimate destination

X.408. Provides recommendations for code and format conversion, such as the conversion between International Alphabet no. 5 (ASCII code) and the S.61 teletex character set.

X.411. Describes the message-transfer layer (MTL) service. This recommendation describes how the MTS user transfers messages with the MTS by defining the service definitions and the abstract syntaxes.

X.413. Contains the provisions for the message store (MS). It describes how MS acts as an intermediary between the UA and the MTS.

X.419. Defines the procedures for accessing the MTS and the MS and for the message exchanges between MTAs. It describes the application-contexts with three MHS protocols, P1, P3, and P7.

X.420. Describes the IPM and the P2 protocol. This service defines the semantics and syntax involved in the receiving and sending of interpersonal traffic. In addition, it recommends the operations for the transfer of the protocol data units through the system.

X.440. Defines a voice messaging application, based on the ITU-T G.721 Recommendation with 32 kbit/s ADPCM.

X.480 through X.485. Define the procedures for conformance testing MHS and directory service (X.500) protocols, including P1, P3, P7, and the voice messaging system protocol.

OSI Realization of MHS

MHS uses the principles of the OSI Reference Model. The entities and protocols reside in the application layer of the model. As we learned earlier, the application entity (AE) consists of the user element (UE) and the supporting application service elements (ASEs). The message handling ASEs can perform two types of service. The *symmetric* service means a UE both supplies and consumes a service; the *asymmetric* service means the UE either consumes or supplies a service but does not do both. The terms *supplier* and *consumer* are used to identify the roles of the two communicating ASEs in providing or consuming a service. A service is made available through a *port*, which is explained in more detail in a later section.

Access to MTS or MS is provided by a number of application service elements:

- Message Transfer Service Element (MTSE): Supports the message transfer submittal functions
- Message Submission Service Element (MSSE): Supports the services of the submission functions

- Message Retrieval Service Element (MRSE): Supports the services of the retrieval functions (for MS)

- Message Delivery Service Element (MDSE): Supports the services of the delivery functions

- Message Administration Service Element (MASE): Supports the services of administrative functions between UAs, MSs, and MTAs and controls subsequent interactions by the means of the four ASEs listed directly above

These ASEs can also be supported by three other ASEs (which are described in chapter 7):

- Remote Operations Service Element (ROSE)
- Reliable Transfer Service Element (RTSE)
- Association Control Service Element (ACSE)

A combination of these eight message handling and supporting ASEs defines the application-context of an application association. Figure 8.5 shows the general relationships of the MHS ASEs and the other ASEs. The specific combinations are defined in the MHS Recommendations.

Remember that an application-context specifies how the association is to be established and which ASEs are used. Every application-context is iden-

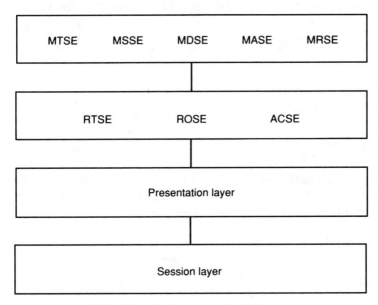

Figure 8.5 The MHS ASEs and other supporting ASEs.

TABLE 8.2 Structure of MHS

Name of recommendation or standard	CCITT	ISO
MHS: System Service and Overview	X.400	10021-1
MHS: Overall Architecture	X.402	10021-2
MHS: Conformance Testing	X.403	
MHS: Abstract Service Definition	X.407	10021-3
MHS: Encoded Information Type Conversion Rules	X.408	
MHS: MTS Abstract Service Definition and Procedures	X.411	10021-4
MHS: MS Abstract Service Definitions	X.413	10021-5
MHS: Protocol Specifications	X.419	10021-6
MHS: Interpersonal Messaging System	X.420	10021-7

tified by an ASN.1 Object Identifier, and the user that initiates an association provides the respondent with this name through the ACSE. The application-context also uses ASN.1 Object Identifiers to identify the application protocol data units (APDUs).

Table 8.2 is a list of the MHS ASEs and their associated functional objects.

Use of ROSE, RTSE, and ACSE

MHS is designed to use a variety of combinations of underlying application layer services (see Figure 8.6). MHS uses ROSE for the request/reply operations. The remote operations of the MTS access protocol (P3) and the MS access protocol (P7) are ROSE class 2 operations. The ASEs are the users of the following ROSE services:

- RO-REJECT-PRO-INVOKE
- RO-RESULT
- RO-ERROR
- RO-REJECT-U
- RO-REJECT-P

If the RTSE is used in the application-context, the MTS and MS are the sole users of the RT-OPEN and RT-CLOSE services of the RTSE. The ROSE is the sole user of the following RTSE services:

- RT-TURN-GIVERT-TRANSFER
- RT-TURN-PLEASE
- RT-TURN-GIVE
- RT-P-ABORT
- RT-U-ABORT

The RTSE may or may not be used in the application-context. If RTSE is not used, the MTS and MS are the sole users of A-ASSOCIATE and A-RE-LEASE services of the ACSE in normal mode. The ROSE is the user of the ACSE A-ABORT and A-P-ABORT services.

If RTSE is used in the application-context, the RTSE is the sole user of these ACSE services: A-ASSOCIATE, A-RELEASE, A-ABORT, A-P-ABORT. The use of the normal mode of the RTSE also implies the use of the normal modes of the ACSE and the presentation layer.

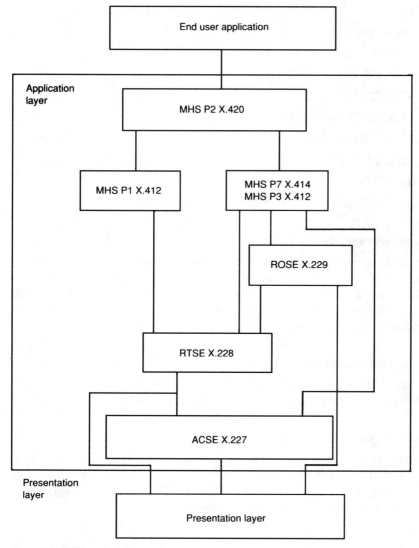

Figure 8.6 MHS underlying services.

Presentation layer services

At the presentation layer the following rules apply. The ACSE is the sole user of the P-CONNECT, P-RELEASE, P-U-ABORT, and P-P-ABORT services. If the RTSE is not included in the application-context, the ROSE is the sole user of the presentation layer P-DATA service. If the RTSE is included in the application-context, the RTSE is the sole user of the following presentation layer services:

- P-U-EXCEPTION-REPORTP-ACTIVITY-START
- P-DATA
- P-MINOR-SYNCHRONIZE
- P-ACTIVITY-END
- P-ACTIVITY-INTERRUPT
- P-ACTIVITY-DISCARD
- P-U-EXCEPTION-REPORT
- P-ACTIVITY-RESUME
- P-P-EXCEPTION-REPORT
- P-TOKEN-PLEASE
- P-CONTROL-GIVE

Session layer services

Regarding the session layer, if the RTSE is used in the application-context, these session level functional units are used by the presentation layer: kernel, half-duplex, exceptions, minor-synchronize, and activity-management. If the RTSE is not used, the presentation layer uses the session level kernel, and duplex functional units.

Transport layer services

The use of transport class 0 is required for transport layer services. Expedited service is not allowed. Other classes are optional.

Lower layer services

MHS does not concern itself with the protocols in the lower three layers of the OSI Model. The MHS message can be transported through X.25 networks, integrated services digital networks (ISDNs), telephone connections, and other types of lower layer systems.

MHS Management Domains

A collection of at least one MTA and zero or more UAs, which are administered by an organization, is called a management domain (MD). The MD is the primary building block of the MHS. An MD controlled by an administration (for example, a Postal, Telephone and Telegraph Agency, or PTT) is designated as an administration management domain (ADMD) and one managed by any other organization is called a private management domain (PRMD).

The basic purpose of the MHS recommendations is to enable the implementation of a global MHS. The ADMDs play the central role in the global MHS. They connect to each other to form the backbone of the global MHS. The PRMDs are attached to the ADMDs and, as such, play a peripheral role in the MHS.

Naming, addressing, and routing in MHS

The key to the management of the global MHS is the conventions for naming and addressing. Every user or DL must have an unambiguous *name*. MHS uses two kinds of names. A primitive name identifies a unique and specific entity such as an employee number or a social security number. A descriptive name denotes one user of the MHS, such as a job title. A descriptive name could identify different entities (in this example, people) as they move through a job. On the other hand, the primitive name is specific to the entity (a person). It may have global uniqueness (unique social security number in the United States) or it may not have global uniqueness (employee number). A name is permitted to have attributes, which further identify an end user (entity) with more detailed parameters.

The MHS *address* specifies the information needed for a message delivery. It can identify the locations of the MHS entities. Typically, a name is looked up in a directory to find the corresponding address, but MHS does allow an originator/recipient (O/R) name to be a directory name, an O/R address, or both. So, it gives the user considerable flexibility in this important area.

MHS addresses consist of *attributes*. An attribute describes a user or a DL, and it may also locate the user or DL within the MHS. An organization has many choices for an attribute list. Generally speaking, the attributes fall under four broad categories (see Table 8.3). Specifically speaking, X.402 defines the standard attributes that comprise an address (see Table 8.4).

The MD is responsible for ensuring that each UA has at least one name, and it must allow users to construct any attributes that are needed by other MDs.

TABLE 8.3 Message Handling ASEs

		Functional objects			
ASE	Form	UA	MS	MTA	AU
MTSE	AS	—	—	C/S	—
MSSE	ASY	C	C/S	S	—
MDSE	ASY	C	C	S	—
MRSE	ASY	C	S	—	—
MASE	ASY	C	C/S	S	—

SY = symmetric
ASY = asymmetric
C = consumer
S = supplier

TABLE 8.4 Typical Examples of O/R Names

Personal attributes	Surname, initials, first name, qualifier (II, Jr.)
Geographical attributes	Street name and number, town, county, region, state, country
Organizational attributes	Name, unit(s) within organization, position or role with organization
Architectural attributes	X.121 address, unique UA identifier, ADMD, PRMD

Security Features in MHS

As the reader might expect, a message system should provide security for the users' traffic. MHS provides for a variety of security service elements. Many of these services rely on the use of encryption/decryption techniques and a separate directory that is very secure. MHS does not define the actual encryption/decryption and directory operations but relies on the use of other OSI services. On a general level, the following security services are available:

- Authentication of the originator of the message
- Authentication of the originator of a nondelivery notice
- Authentication that the message was delivered, that its contents were not altered, and that all recipients received a copy
- Proof that the message was not altered
- Measures to prevent the examination of the message
- Authentication of correct message sequencing
- Inability of a recipient to claim the message was not received
- Inability of sender to claim the message was sent, when it was not submitted to MHS. (No longer can we say, "The check is in the mail!")

The P Protocols

This section examines each of the MHS P protocols. Without delving into ASN.1 code, we must restrict the discussion to the architecture of the P protocols and trust the reader will gain a general understanding of them. Figure 8.7 provides a view of these protocols in relation to the major MHS components. Like many of the newer ITU-T recommendations, MHS is modeled as a number of objects. The internal structures of these objects are not described, but the services offered by the objects are. The services that are offered to a user are made available through *ports*. A port represents a view of the services provided by the MHS objects. Figure 8.7 shows the ports that are associated with each of the P protocols. Ports may be symmetrical or asymmetrical in relation to the provision or consumption of services.

P3-MTA user and MTA

The MTA user to MTA service (P3) is organized as shown in Figure 8.8 to obtain the MTS. It consists of three ports: submission, delivery, and administration.

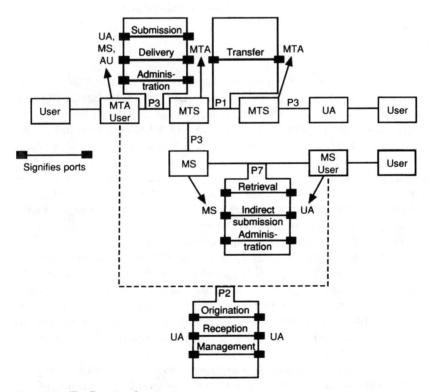

Figure 8.7 The P protocol ports.

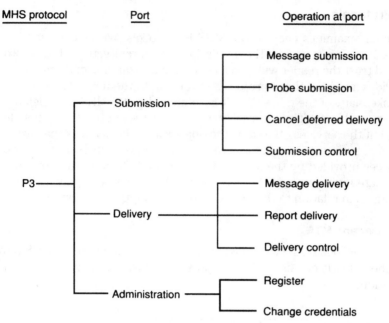

MHS protocol   Port   Operation at port

Figure 8.8 The P3 ports and operations.

The *message submission port* is organized around the following abstract operations: (1) message submission, (2) probe submission, (3) cancel deferred delivery, and (4) submission control. The purpose of the
message submission operation is to permit an MTA user to transmit a message to the MTA. This message must contain at least the originator name;
the recipient name; the type of message requested by the message originator (if any); the content type of the message, such as an interpersonal message; and the actual contents of the message, which is not examined by the
MTA module. Many other parameters can be placed in the message during
the message submission operation. The reader should refer to X.411 for a
detailed description of these other optional arguments.

The *probe submission* operation allows an MTS user to send a probe to
the MTS to determine if the potential message could be transferred and delivered without problems. This message also contains the originator name,
the recipient name, the content type of the potential message to be transmitted, and the type of report requested by the message originator. Of
course, a large number of arguments are optional and may be included in
this operation as well.

The *cancel deferred delivery* operation allows an MTS user to request
the MTS to cancel a message that was previously submitted for deferred delivery to the recipient. The MTS user must give to the MTS the message

submission identifier. This identifier is also in the message whose deferred delivery is to be canceled.

The *submission control* operation is used by the MTS to limit the MTS user in the flow of messages the user may submit to the MTS. With this operation, the MTS is allowed to define what operations are permitted between the MTS and the MTS user, to describe priority thresholds in which messages may be submitted, to define a longest allowable length for a message that might be submitted, and to provide guidelines in relation to some of the security operations that may be supported.

The *delivery port* consists of the following operations: (1) message delivery, (2) report delivery, and (3) delivery control.

The *message delivery* operation is used by the MTS to deliver a message to the MTS user. At a minimum, this message must contain a message delivery identifier, the time of the message delivery, a time at which the MTS has accepted responsibility for the message, the O/R name of the recipient of the message (generated by the MTS), the content type of the message, and, of course, the content. As the reader might expect, a large number of other parameters can be contained in this message.

The *report delivery* service is used by the MTS to provide an acknowledgment to the MTS user. The acknowledgment depends on previous submissions to the MTS, such as a message submission operation or a probe submission operation. The MTS must return this message with a subject submission identifier, which identifies the submitter of the message. It must also convey back the actual recipient name, which is the O/R name of the message recipient. This is generated by the originator of the message or, if the message has been redirected, by the MTS.

MHS has very powerful features for providing status reports between the various entities in the system. For example, in the *report delivery* operation, the MTS may generate up to 49 diagnostic codes explaining the reason for delivery problems or probing problems. Diagnostic indicators such as an invalid name, congestion at the MTS, invalid arguments, designation for too many recipients, coding problems, recipient reassignment problems, expansion list problems, etc., are a few of the report delivery status indicators.

The *delivery control* operation allows the MTS user to limit the MTS in sending messages to the user. If this operation is invoked, the MTS has the option of holding the messages or releasing them. The operation provides several capabilities for the MTS user to flow control the MTS. Content types, priority identifiers, maximum length of traffic, security contexts, etc., can all be invoked to support this service.

The last port for the MTS user to MTS operation is the *administration port*. It consists of the register operation and the change credentials operation.

The *register* operation allows the MTS user to make changes to its profile with the MTS in relation to message delivery. The register may be altered by yet another register operation. This message contains the O/R name of the MTS user, the user address, and the information relating to the profiles that are to be changed. The number of parameters to be changed in the profile are quite diverse, giving the MTS user considerable control on its services with an MHS.

Finally, the *change credentials* operation allows the MTS user to change the user's credentials that are held at the MTS. It also allows the MTS to change credentials held by the MTS user. These credentials are used for authentication purposes. The only parameters required for this message are the old credentials' and the new credentials' values.

P1-MTA and MTA

The MTA to MTA service (P1) is used within the MTS and is organized as shown in Figure 8.9. The P1 protocol defines one port. This port is called the *transfer port* and supports three operations: (1) message transfer, (2) probe transfer, and (3) report transfer.

The *message transfer* operation is used to transfer messages from one MTA to another. As with the other protocols in MHS, a large number of optional fields can be carried in the message and some fields are required. The message identifier must be in each message; it is used to identify the message and distinguish it from other messages in the MTS. It is generated by the originating MTA and it must have the same value as the message submission identifier supplied by the originator of the message (typically a UA). For multiple copies to different MTAs, each copy must contain this identifier. Trace information is also required in order to determine operations relating to probes or reports. The originator name must be in all messages as must be the recipients' names. If multiple copies are sent, a message identifier is also used to determine which copy of the message is being received by the recipient. A responsibility field is also contained in the message; it determines if the receiving MTA must assume the responsibility to deliver the message to the end recipient or relay it to yet another MTA.

The message also contains an originating MTA report request. This field is used by the MTS to determine what kind of report is requested by the

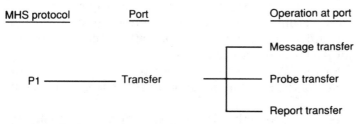

Figure 8.9 The P1 port and operations.

originating MTA. It is used to determine if a report is to be returned to the originator in case of nondelivery of the message. It is also used to indicate if trace information is desired for delivery in the event of delivery problems. When used in conjunction with another required field (the originator report request field), it determines if the report may be suppressed.

As the reader might expect, this message also contains the encoded information type field as well as the contents of the message.

The *probe transfer* and *report transfer* operations contain information similar to that in the MT operation just discussed. In addition, the probe transfer contains additional fields to provide information between the MTAs participating in the probe. The report transfer operation allows status information to be transferred between MTAs. These two messages contain information such as trace information, originally specified recipient numbers, distribution lists, expansion history, etc.

P7-MS user and MS

When the reader examines the MS protocol, it will be evident that many of the operations at the MS simply mirror the operations at the MTA. Indeed, as Figures 8.3 and 8.7 indicate, the MS is acting as an intermediary between the UA and the MTA. Thus, the protocol is quite similar to many of the operations found in the MTA operations discussed in P3.

The MS user to MS service (P7) is organized as shown in Figure 8.10. The P7 protocol defines three ports: the retrieval port, the indirect-submission port, and the administration port.

The *retrieval port* specifies the following six operations: (1) summarize, (2) list, (3) fetch, (4) delete, (5) register MS, and (6) alert.

The *summarize* operation is used to return summary information from a database to a user (a UA). The type of information that is retrieved by this operation is not defined by MHS. Rather, ASN.1 arguments are used to tailor the summarize operation to the specific needs of the user. This operation also allows the creation of searches based on ASN.1 filters. MS makes several checks on the UA request (security, permissible search, etc.) before passing the information to the UA.

The *list* operation allows the search of an MHS database for items that may be of interest to a user. Again, the list operation is tailored through the use of arguments to ASN.1 macros.

The retrieval port's *fetch* operation provides for more specific services than the summarize and list operations just discussed. It is used to obtain information about a specific entry in an MHS database. Through the use of ASN.1 arguments, the identification of the information to be retrieved is provided to MS, which then makes the necessary fetch.

In any store-and-forward operation, it is important to delete entries from the MS database; otherwise the arriving messages will accumulate and create

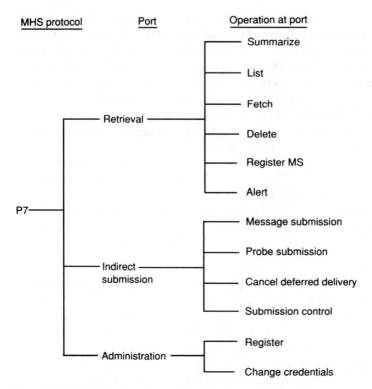

Figure 8.10 The P7 ports and operations.

storage problems. Therefore, P7 supports a *delete* operation that allows a user to delete selected entries from the database. It allows linked operations where dependent child entries may also be deleted, although specific rules in this protocol define how dependent entries are allowed to be deleted.

The *register-MS* operation is used to establish (register) or disestablish (deregister) information at the message store. The entries that are registered or deregistered are a function of the parameters in the ASN.1 macros. This operation is important for establishing new and old credentials for security measures, as well as changing security labels. Be aware that this operation is not restricted to security aspects, but may be used for anything that users wish to register and deregister with MS. The MS returns a message to the UA that indicates if the operation is successful.

The final operation for the MS P7 retrieval port is the *alert* operation. It is used by MS to immediately inform the MS user of a new arrival at the MS. The alerting is actually determined by the registration of alert selection criteria.

The *indirect submission* port of P7 consists of the following operations: (1) message submission, (2) probe submission, (3) canceled defer delivery, and (4) submission control.

The *message submission* operation is performed by the MS to ensure that the information in the message it receives is valid. It also determines if the message content was supplied by the user agent or if it must be inserted by the MS. If an association does not exist between the MS and the MTA, it establishes an association before sending traffic to the MTA. It then invokes a message transmission operation (discussed in protocol P3) and waits for a message submission result from the MTA.

The *probe submission* operation is used by the MS to support a probe operation. Upon receiving a probe submission message, it checks to see if the fields in the message are valid. It also makes certain that an association exists between the MS and the MTA. If all goes well, it then submits a probe submission to the MTA and awaits a probe submission result (again, defined in protocol P3).

The *cancel deferred delivery* operation is used by the operation to support the P3 canceled deferred delivery operation. The MS sends the canceled defer delivery message to the MTA and awaits the return of the canceled deferred delivery result.

The MS is also involved in the *submission control* operation. Again, it is simply serving as an intermediary to support the P3 submission control operation, which was discussed earlier.

The *administration* port of P7 includes two operations: (1) *register* and (2) *change credentials*. As the reader might expect, these operations are used by the MS to support the MTAs (P3) administration port and the registration and change credentials operations discussed earlier.

P2-IPM-UA to IPM-UA

The IPM-UA to IPM-UA (P2) is organized as shown in Figure 8.11. The IPM consists of three ports: origination, reception, and management.

The *origination* port is used by a user to convey the interpersonal messaging system messages. It consists of the following operations: (1) originate probe, (2) originate IPM, and (3) originate receipt notification (RN).

The *originate probe* operation allows an end user to originate a probe about a class of messages. The UA supplies the probe information to the MTS except for the O/R names of the recipients, the options for each recipient, and some special options.

The *originate IPM* is used by the end user to send a message to the MHS. The end user need only furnish the recipient names and any per-message options such as delivery instructions, priority, etc.

The *originate RN* operation allows an interpersonal notification (IPN) to report an originator's receipt of an IPM.

The *reception* port consists of the following operations: (1) receive report, (2) receive IPM, (3) receive RN, and (4) receive nonreceipt notification (NRN).

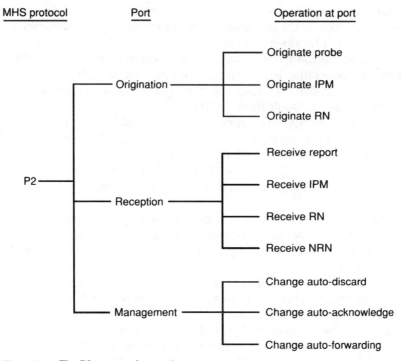

| MHS protocol | Port | Operation at port |

Origination:
- Originate probe
- Originate IPM
- Originate RN

Reception:
- Receive report
- Receive IPM
- Receive RN
- Receive NRN

Management:
- Change auto-discard
- Change auto-acknowledge
- Change auto-forwarding

P2

Figure 8.11 The P2 ports and operations.

The *receive report* operation provides reports about the status of IPM operations such as probes, auto-discards, auto-forwards, etc. The *receive IPM* defines the procedures for receiving an IPM. The *receive RN* and *receive NRN* operations define the procedures for receiving IPMs whose content is either an RN or an NRN.

The *management* port of P2 consists of the following operations: (1) *change auto-discard,* (2) *change auto-acknowledgment,* and (3) *change auto-forwarding.* These operations are used to alter the profiles for automatically discarding, acknowledging, and forwarding messages.

The *reception* port is used by the IPMs to send messages to the end user. A reception port is established for each user for the IPM. This port consists of the following operations: (1) receive report, (2) receive IPM, (3) receive RN, and (4) receive NRN.

The *management* port is used by the user of an IPM service to obtain information about capabilities the IPM provides the user such as acknowledgment, forwarding, and automatic discarding of traffic, etc. The port consists of the following operations: (1) *change auto-discard,* (2) *change auto-acknowledgment,* and (3) *change auto-forwarding.*

MHS Service Elements

Several references have been made to service elements. They represent the "heart" of the MHS recommendations because they define the features and functions of the system. Figure 8.12 shows the structure of the service elements.

Because the service elements are quite important to MHS, it is appropriate to examine them in some detail. The full list of the service elements is provided in Table 8.6, and a short paragraph describes each. The table specifies if the service element is part of the 1984 or 1988 ITU-T Recommendations.

MHS service elements' descriptions

We now examine each of the MHS service elements. An examination of these elements shows the richness of functions inherent in the MHS. At the same time, it is recognized that this section may be a bit tedious because each service element is described, and there are many elements, as evidenced in Table 8.6. Therefore, the reader may wish to scan through this section and read the descriptions of the service elements that might be useful with a particular application.

To facilitate the comparison of the entries in this section to the list of service elements in Table 8.6, most of the service elements are in alphabetical order, unless they are described with other complementary service elements.

The *access management* service element provides the support for the UA and MTA to communicate with each other through the identification and validation of names and addresses. This service element allows the UA to use its O/R name for access security. The MTS passwords used for this

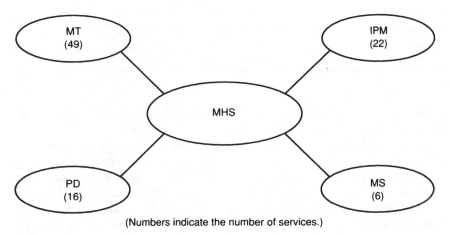

(Numbers indicate the number of services.)

Figure 8.12 The MHS service elements.

TABLE 8.5 Standard Attributes

Attribute type	Attribute description
Administration-domain-name	ADMD relative to a country name
Common-name	Identification such as an organization or job title
Country-name	Country identifier with X.121 or ISO 3166
Extension of postal O/R-address address (organizational unit)	Additional identifier within postal components
Extension physical-delivery-address-components	Additional identifier within postal address (room and floor numbers)
Local-postal-attributes	Identifies locus of distribution (geographical area)
Network address	X.121, E.163, E.164, PSAP address
Numeric-user-identifier	Identifies user relative to the ADMD
Organization-name	Identifies an organization
Organizational-unit-names	Units of an organization (e.g., division)
Physical-delivery-service-name	Identifies physical delivery relative to the ADMD
Personal name	A person's name
Physical-delivery-country-name	Country in which user takes delivery
Physical-delivery-office-number	Distinguishes among several post offices
Physical-delivery-organization-name	A postal patron's organization
Physical-delivery-personal-name	A postal patron
Post-office-box-address	Number of post office box
Postal-code	A geographical area postal code
Poste-restante-address	Identifies code given to post office collection of messages for user
Private-domain-name	Identifies a PRMD
Street-address	Identifies a street address
Terminal-identifier	Identifier of a terminal
Terminal-type	Type of terminal (telex, FAX, etc.)
Unformatted-postal-address	Postal address in free form
Unique-postal-name	Identifies a unique postal name (a "plaza," hamlet, etc.)

service element are different from any of those used by the UA itself to authenticate the end user. If the MHS user wishes a more secure form of access, the secure access management service element can be used.

The *additional physical rendition* allows the originating MHS user to request the kind of physical rendition that is to be done on the message. Physical rendition is the transformation of the electronic message to a physical message, such as the type of paper, the kind of printing, stipulation for color rendition, etc. This service is performed by the physical delivery access unit (PDAU). Use of this service element requires bilateral agreement.

TABLE 8.6 MHS Service Elements

Service element	MT	IPM	PD	MS
Access management	84	—	—	—
Additional physical rendition	—	—	88	—
Alternate recipient allowed	84	—	—	—
Alternate recipient assigment	84	—	—	—
Authorizing users indication	—	84	—	—
Auto-forwarded indication	—	84	—	—
Basic physical rendition	—	—	88	—
Blind copy recipient indication	—	84	—	—
Body part encryption indication	—	84	—	—
Content confidentiality	88	—	—	—
Content integrity	88	—	—	—
Content type indication	84	—	—	—
Conversion prohibition (CP)	84	—	—	—
CP in case of information loss	88	—	—	—
Conversion indication	84	—	—	—
Counter collection	—	—	88	—
Counter collection with advice	—	—	88	—
Cross-referencing indication	—	84	—	—
Deferred delivery	84	—	—	—
Deferred delivery cancellation	84	—	—	—
Delivery notification	84	—	—	—
Delivery time stamp indication	84	—	—	—
Delivery via Bureaufax Service	—	—	88	—
Designation of recipient by directory name	88	—	—	—
Disclosure of other recipients	84	—	—	—
DL expansion history indication	88	—	—	—
DL expansion prohibited	88	—	—	—
Express mail service (EMS)	—	—	88	—
Expiry date indication	—	84	—	—
Explicit conversion	84	—	—	—
Forwarded IP-message indication	—	84	—	—
Grade of delivery selection	84	—	—	—
Hold for delivery	84	—	—	—
Implicit conversion	84	—	—	—
Importance indication	—	84	—	—
Incomplete copy indication	—	88	—	—
IP-message identification	—	84	—	—
Language identification	—	88	—	—
Latest delivery designation	88	—	—	—

TABLE 8.6   MHS Service Elements (Continued)

Service element	MT	IPM	PD	MS
Message flow confidentiality	88	—	—	—
Message identification	84	—	—	—
Message origin authentication	88	—	—	—
Message security labeling	88	—	—	—
Message sequence integrity	88	—	—	—
Multidestination delivery	84	—	—	—
Multipart body	—	84	—	—
Nondelivery notification	84	—	—	—
Nonreceipt notification request indication	—	84	—	—
Nonrepudiation of delivery	88	—	—	—
Nonrepudiation of origin	88	—	—	—
Nonrepudiation of submission	88	—	—	—
Obsoleting indication	—	84	—	—
Ordinary mail	—	—	88	—
Original encoded information type indication	84	—	—	—
Originator indication	—	84	—	—
Originator requested alternate recipient	88	—	—	—
Physical delivery notification by MHS	—	—	88	—
Physical delivery notification by PDS	—	—	88	—
Physical forwarding allowed	—	—	88	—
Physical forwarding prohibited	—	—	88	—
Prevention of nondelivery notification	84	—	—	—
Primary and copy recipients indication	—	84	—	—
Probe	84	—	—	—
Probe origin authentication	88	—	—	—
Proof of delivery	88	—	—	—
Proof of submission	88	—	—	—
Receipt notification request indication	—	84	—	—
Redirection disallowed by originator	88	—	—	—
Redirection of incoming messages	88	—	—	—
Registered mail	—	—	88	—
Registered mail to addressee in person	—	—	88	—
Reply request indication	—	84	—	—
Replying IP-message indication	—	84	—	—
Report origin authentication	88	—	—	—
Request for forwarding address	—	—	88	—
Requested delivery method	88	—	—	—
Restricted delivery	88	—	—	—
Return of content	84	—	—	—
Secure access management	88	—	—	—

TABLE 8.6 MHS Service Elements (Continued)

Sensitivity indication	—	84	—	—
Special delivery	—	—	88	—
Stored message alert	—	—	—	88
Stored message auto-forward	—	—	—	88
Stored message deletion	—	—	—	88
Stored message fetching	—	—	—	88
Stored message listing	—	—	—	88
Stored message summary	—	—	—	88
Subject indication	—	84	—	—
Submission time stamp indication	84	—	—	—
Typed body	—	84	—	—
Undeliverable mail with return of message	—	—	88	—
Use of distribution list	88	—	—	—
User/UA capabilities registration	88	—	—	—

The *alternate recipient allowed* service element allows the originating UA to specify the delivery to an alternate recipient. The destination MD uses the UA's attributes to select the recipient UA.

The *alternate recipient assignment* service element gives the UA the capability to have certain messages delivered to it even when there is not an exact match in the attributes or descriptive names of the UA. In this situation, the UA is specified in specific attributes and also in other attributes for which any value would be acceptable. This is useful when a message is sent that contains the proper domain name and perhaps company name, but the person's name in the message does not match. This means the UA would still have the message delivered to it. The message will not be rejected, but manual procedures could be implemented to handle the notification to the person or to locate the individual. Typically, certain attributes must match (such as company name) and other attributes need not match (such as a person's name).

The *authorizing users indication* service element allows the originator to identify the names of one or more persons who have authorized the sending of the message. The actual purpose of this service element is to give the receivers the identification of the people who are authorized to send the traffic, in contrast to the person who sent it. The latter person is the originator; the former is the authorizer.

The *auto-forwarded indication* service element is used to determine that an incoming IP-message contains a message that has been auto-forwarded. The auto-forwarded IP-message may be accompanied by the information that was intended with its original delivery, such as code conversion indications and time stamps.

The *basic physical rendition* service element allows the PDAU to provide the conversion of the MHS message to actual physical copy, with specified physical characteristics, such as the color of paper, etc.

The *blind copy recipient indication* service element allows the originator of the message to provide the O/R names of additional recipients of the traffic. These names are not disclosed to any of the other recipients. DLs can also be blind copy recipients.

The *body part encryption indication* service element is used by the originator to inform the receiver that the body part (or a portion thereof) in the IP-message has been encrypted. In turn, it can then be used by the recipient to determine which part of the message must be decrypted. The service element does not perform any encryption or decryption.

The *content confidentiality* service element is used by the originator of the message to prevent the disclosure of the message to unauthorized recipients. This service element is used on a per-message basis and can use asymmetric or symmetric encryption.

The *content integrity* service element is used by the originator to give the recipient the means to determine that the content of the message has not been modified. The service element is used on a per-message basis and can use asymmetric or symmetric encryption.

The *content type indication* service element supports an originating UA's indication of the content type (syntax and semantics) for each message. The recipient UA may have one or more content types delivered to it. The content type identification specifies the particular type.

The *conversion prohibition* service element is used by the originating UA to instruct the MTS that the message is not to have any implicit encoded information type conversion performed on it (for example, IA5 code to group 3 FAX).

Four other service elements in this category provide complementary services. The *implicit conversion* service element instructs the MTS to perform conversion for a period of time. It is so named because the UAs are not required to explicitly request the service. When the conversion is performed and the delivery made, the recipient is informed of the original and current encoded information types.

In contrast to the implicit conversion service element, the *explicit conversion* service element is used by the originating UA to instruct the MTS to perform specific conversion services. When using this service element, the receiving UA is informed of the original code types as well as the newly encoded types of the message.

The *conversion prohibition in case of loss of information* service element is used by the originating UA to direct the MTS not to perform code conversion if it might possibly result in the loss of information. This service element could be used if, for example, the characteristics of the recipient's I/O device are not known.

The *converted indication* service element is used by the MTS to inform a recipient UA that the MTS performed code conversion on a delivered message. The service element allows the recipient UA to be informed of the resulting code types.

The *counter collection* service element is used by the originator to instruct the physical delivery system (PDS, like a postal service) to hold the physical message for pickup by the recipient.

The *counter collection with advice* service element allows the originator to inform the PDS to hold the message and to inform the recipient that the message is available for collection. The originator provides the identification (telephone number, etc.) of the recipient.

The *cross-referencing indication* service element is used to identify associated IP-messages. This is useful to allow the IPM UA to retrieve copies of the referenced messages.

The *deferred delivery* service element is used by the originating UA to inform the MTS that the message is to be delivered within a specified date and time. The MTS attempts to deliver the traffic as close to this period as possible. It will not deliver the traffic before the deferred delivery time and date stamp. The originator's MD may place a limit on the parameters in the service element.

The *deferred delivery cancellation* service element is used by the originating UA to inform the MTS that it is to cancel a message that was previously submitted. The cancellation may or may not take effect. For example, if the message has already been forwarded within the MTS, it will not be canceled.

The *delivery notification* service element is used for an end-to-end acknowledgment. This allows the originating UA to request that it be informed when the message has been successfully delivered to the recipient UA or the access unit. This service element also stipulates the time and date of the actual delivery. The service element does not mean that the recipient UA has acted on the message or has even examined it. It only states that the delivery did occur. It is similar in concept to receiving a receipt of a delivery of a letter through the postal service. It merely means that the recipient has signed for the letter; it does not mean the recipient read the letter. The originator may also request this service element on a multidestination message on a per-recipient basis.

The *delivery time stamp indication* service element is used by the MTS to inform the recipient UA of the time and date the MTS delivered the message. If the message is delivered to a PDAU, this service element indicates the time and date the PDAU has taken responsibility for the message.

The *delivery via Bureaufax Service* element is used by the originator to direct the PDAU and the PDS to use the Bureaufax Service for the delivery of the message.

The *designation of recipient by directory name* allows the originating UA to use a directory name instead of the recipient's O/R address.

The *disclosure of other recipients* service element allows the originating UA to require the MTS to disclose the O/R names of all the other recipients after they have received a copy of a multirecipient message. These names are originally supplied by the originating UA, and the MTS then informs the UA when each recipient receives the message.

The *DL expansion history indication* service element gives the recipient information about the DLs used to affect the message delivery.

The *DL expansion prohibited* service element is used if the originator does not wish a recipient to refer to a distribution list. If so, no DL expansion can occur and a nondelivery notification is returned to the originating UA.

The *express mail service (EMS)* element is used by the originator to direct the PDS to deliver the physical message through an express mail service.

The *expiry date indication* service element is used by the originator to indicate to the recipient the date and time that the IP-message will no longer be valid. However, MHS does not specify what happens in the event the expiration time and date are exceeded.

The *forwarded IP-message indication* service element is used to forward the IP-message as the body of the IP-message. In a multipart body, the forwarded parts can be included along with the message body parts of other types of traffic. The message indicates that the body part contains a forwarded IP-message.

The *grade of delivery selection* service element allows the sending UA to request the MTS to transfer the message on an urgent or nonurgent basis, rather than on a normal basis. The recipient of the message also receives notice about the grade of delivery. The service element does not stipulate the specific times for the grade of delivery selection.

The *hold for delivery* service element is quite similar to the postal department's practice of holding mail for individuals. This allows the UA to request the MTS to hold its traffic for delivery as well as for any returning notifications. The UA can use this service element to inform the MTS when it is unavailable and when it is again ready to accept traffic, although MHS stipulates that the criteria for requesting this service element are restricted to the following: (1) encoded information type, (2) content type, (3) maximum content length, and (4) priority.

The hold for delivery service element is a temporary storage facility only. It should not be confused with the MHS MS service. The MS is intended to provide storage for an extended period of time.

The *implicit conversion* service element is discussed in context with the conversion prohibition service element.

The *importance indication* service element is used to establish a priority for the IP-message. MHS defines three levels of priority: low, normal, and high. While this service element is used to indicate how important the traffic is, it is not related to the delivery selection service provided by the MTS itself. MHS does not specify how to handle this service element. This service element is not related to the grade of delivery service element, which is discussed later.

The *incomplete copy indication* is used by the originator to indicate the IP-message. It also indicates that one or more body parts or heading fields are missing.

The *IP-message identification* is used by IPM UAs to uniquely identify each IPM message. It consists of an O/R name or the originator and a value unique to that name.

The *language indication* service element allows the originating UA to indicate which languages are in the IP-message.

The *latest delivery* service element is used by the originating UA to stipulate the latest time the message can be delivered. If the MTS cannot meet this time, the message is canceled and not delivered. If a message is destined for multiple recipients, a time expiration will affect only those recipients that have not yet received the message.

The *message flow confidentiality* service element is used by the originator to prevent an interpretation of the message contents by observing the message flow. The 1988 MHS release does not fully support this service element.

The *message identification* service element is used by the MTS to give the UA the unique identification for each message or probe sent to or from the MTS. The UAs and MTs also use these values to identify previously submitted traffic of other service elements, such as confirmation and nondelivery notification.

The *message origin authentication* service element provides the message recipient(s) and any MTA information on the authentication of the origin of the message. This service can be provided on a per-message or per-recipient basis, and it uses encryption techniques.

The *message security labeling* service element is used by the originator to identify the sensitivity of the message through the use of a security label. The label is used to determine how to handle the message. MHS does not define the handling procedures.

The *message sequence integrity* service element allows the recipient to determine that messages have arrived from the originator in the proper order.

The *multidestination delivery* service element is used by the originating UA to stipulate that a message is to be submitted to more than one receiving UA. The specification does not place a limit on the number of UAs that can receive the message, nor does it place rules on simultaneous delivery.

The *multipart body* service element is used by the originator to send to the recipient an IP-message that is in several parts. Each part contains the nature, attributes, and type of the message.

The *nondelivery notification* service element is used by the MTS to inform an originating O/R if a message was not delivered to the receiving UA. The reason for the nondelivery is indicated as part of the service element.

This service element supports multidestination messages and distribution list expansions.

The *nonreceipt notification* service element is used by the originator to request that it be notified if an IP-message is not delivered. This can be requested on an end recipient basis for a multidestination message. The receiving IPM UA must return a nonreceipt if the message was forwarded to another recipient, if the recipient did not subscribe to the service, or if the message was discarded before the reception actually occurred. The nonreceipt notification must contain the identification of the IP-message involved in the nondelivery, as well as reasons for the nondelivery.

The *nonrepudiation of delivery* service element gives the message originator absolute proof that the message was delivered to the recipient. It eliminates the possibility that the recipient can claim the message never arrived. The service element is provided using asymmetric encryption techniques.

The *nonrepudiation of origin* service element allows the recipient of a message to have absolute proof of the origin of the message. It prevents the originator from denying to have been the originator. It also prevents the originator from revoking the message. The service element is provided using asymmetric encoding techniques.

The *nonrepudiation of submission* service element is similar to the nonrepudiation of delivery, except it pertains to the MTS. In other words, it prevents the MTS from denying that the message was delivered to it.

The *obsoleting indication* service element is used by the originator to indicate that one (or more) IP-messages previously sent is (are) no longer valid. The IP-message identifying this situation supersedes the previous IP-message(s). This MHS service is a useful tool for purging old and/or obsolete messages from a file.

The *ordinary mail* service element allows the PDS to deliver the letter produced by the MHS through the postal service.

The *original encoded information type indication* service element is used by the sending UA to identify the code type of the message submitted to the MTS. When the message is delivered to the recipient UA, it gives this UA the information about the information code.

The *originator indication* service element is used to identify the originator of the traffic. It is the responsibility of the MTS to provide the proper O/R address or directory name to the receiver. MHS states that the originator can be identified to the MTS in a user-friendly way.

The *originator requested alternate recipient* service element allows the originating UA to designate one alternate recipient for each intended recipient. If the delivery cannot be made to the intended recipient, the message is passed the alternate. The alternate may be a DL. MHS considers the delivery of the message to the alternate recipient to be the same as a delivery to the intended recipient in the matters of delivery and nondelivery notifications.

The *physical delivery notification by MHS* service element is used by the originator to request the MHS to notify the originator of the successful or unsuccessful delivery of the physical message. The PDS does not provide a physical record of this notification.

The *physical delivery notification by PDS* is similar to the above service element, except the PDS provides the notification and a physical record for the originator.

The *physical forwarding allowed* service element is used by the PDS to send a physical message to the forwarding address, and the recipient has made this address known to the PDS.

The *physical forwarding prohibited* service element allows the originator to prevent a message from being forwarded to a forwarding address.

The *prevention of nondelivery notification* service element is used by the sending UA to inform the MTS not to return a nondelivery notification if the message is not delivered. This situation is not unusual in electronic mail when nonpriority bulletins are sent to various recipients. The service can be requested on a per-recipient basis for a multidestination message.

The *primary and copy recipients indication* service element is used by the originator to provide the names of the users and DLs who are the primary recipients of the IP-message as well as the identification of the individuals and DLs who are copy recipients only. This is the familiar "to" and "CC" identifiers of a typical letter.

The *probe* service element is used by the sending UA to determine if a message can be delivered before it is actually sent. The MTS is responsible for providing the submission information by generating either nondelivery or delivery notifications. The useful aspect of the probe service element relates to the possibility that different codes, different message sizes, or even different content types might not be deliverable. In such a case, the originating UA might alter some of its size, content types, and codes as a result of the probe. If the probe is based on a DL, it only indicates that the originator has the right to submit the DL; it has nothing to do with the likelihood of success.

The *probe origin authentication* service element is used to authenticate the origin of the probe. This service uses the asymmetric encryption technique.

The *proof of delivery* service element is used to authenticate (1) the identity of the recipients and (2) the content of the delivered message. In a sense, it is similar to the practice of message echo used in some lower layer protocols. The service uses symmetric or asymmetric encryption techniques.

The *proof of submission* service element is used by the originator as an assurance by the MTS that the message was submitted for delivery to the recipient.

The *receipt notification request indication* service element is used by the originator to request that it be notified of the receipt of the IP-message.

The receipt notification must include the identifier of the message to which the notification is applicable, the time of destination receipt, the O/R name of the recipient, and an indication if any code conversion was performed.

The *redirection disallowed by originator* is used to override the redirection of incoming messages service element. In this situation, the message will not be redirected, even though the recipient has so requested.

The *redirection of incoming messages* service element is used by a UA to request the MTS to redirect messages to another UA or a DL. The service can be established for a specified period of time, or it can be revoked at any time.

The *registered mail* service element is used by the originator to direct the PDS to handle the physical message as registered mail.

The *registered mail to addressee in person* service element is used by the originator to direct the PDS to handle the message as described above and deliver it only to the addressee.

The *reply request indication* service element is used by the originator to request the receiver send an IP-message and reply to the original message. This service element also allows the originator to specify the date on which to send the reply and the O/R names of those who should receive the reply. The recipient is responsible for deciding whether or not to reply to the traffic.

The *replying IP-message indication* service element is used to indicate that the IP-message is sent in reply to another IP-message. The reply may be sent only to the originator or to other users who received copies of the message. It can also be sent to a DL.

The *report origin authentication* service element is used by the message or probe originator to authenticate the origin of a report on the delivery or nondelivery of a message or probe. The service uses the asymmetric encryption technique.

The *request for forwarding address* service element is used by the originator to direct the PDS to give the originator the forwarding address for the message recipient, if this recipient has so indicated to the PDS. This service can be used in conjunction with the physical forwarding allowed and prohibited service elements.

The *requested delivery method* service element is used to request a method of preference for the message delivery. For example, an AU or PDAU can be designated.

The *restricted delivery* service element is used by the recipient UA to inform the MTS that it cannot accept messages from designated UAs or DLs. These entities become unauthorized originators.

The *return of content* service element is used by the sending UA to request that the content of its message be returned in the event of nondelivery. The return will not occur if the system has encoded the message into a different code.

The *secure access management* service element is an MHS security feature. It allows the establishment associations between MTS users, MTSs, or

MTAs. It supports the establishment of the association through strong credentials.

The *sensitivity indication* service element is used by the originator to provide guidelines about the security of the IP-message. This service element indicates whether the recipient should have to identify itself before it receives the traffic, whether the IP-message can be printed on a shared device, and whether the IP-message can be auto-forwarded. The service specifies three levels of sensitivity: (1) personal (the IP-message is sent to an individual recipient, not based on the role of the individual in the organization), (2) private (the IP-message contains information that is to be used only by the recipient and no one else), and (3) company confidential (the IP-message contains information that should only be used according to company procedures).

The *special delivery* service element is used by the originator to direct the PDS to send the physical message through the postal service special delivery service.

The *stored message alert* service element allows a user to specify to its MS a set of criteria that are used to cause the MS to send an alert to the user. The alert is sent to the user through the user's UA.

The *stored message auto-forward* service element requests the MS to automatically forward certain messages. The MHS user registers with the MS the criteria to be used to determine if the message is to be forwarded. The messages that meet a set of criteria are forwarded to one or more users or a DL.

The *stored message deletion* service element is used by the recipient UA to delete messages from the MS queues. Messages cannot be deleted until they have been listed to the UA.

The *stored message fetching* service element is used by the recipient UA to retrieve a message or a portion of a message. The fetching is based on a set of criteria obtained from the use of the stored message listing service element.

The *stored message listing* service element is used by the MS to provide the recipient UA with a list of messages at the MS and the attributes of the messages. The UA can limit the number of messages listed to it.

The *stored message summary* service element is used by the UA to request the MS to provide a count of the number of messages stored at the MS. The count is based on selection criteria (attributes) chosen by the UA.

The *subject indication* service element is used for the originator to indicate to the receiver the subject matter of the IP-message.

The *submission time stamp indication* service element is used by the MTS to inform the sending UA of the date and time the message was submitted to the MTS.

The *typed body* service element defines the attributes of the body of the IP-message. For example, it may describe the type code (IA5), or whether it is a facsimile or a teletext document. These descriptions are carried in the message along with the body.

The *undeliverable mail with return of physical message* service element is used by the PDS if it cannot deliver the physical message. The message is returned to the originator with a reason for the nondelivery.

The *use of DL* service element allows the originating UA to use a DL in place of a number of individual recipients. These recipients can be users or DLs.

The *user/UA capabilities registration* service element allows the UA to inform its MTA of the nature of the messages it wishes to have delivered to it. Delivery restrictions can be based on (1) content type of message, (2) length of message, and (3) encoded information type of message.

Basic and optional MHS services

The use of all these service elements in a system is usually unnecessary. Most users do not need all the services, and a full-feature implementation requires the writing of a considerable amount of code. Like several of the other layers and entities of OSI, the MHS is divided into "required" and optional services. This approach permits a developer to implement a subset of MHS. The required services are called basic (or base) services and are listed in Table 8.7. The other service elements are optional.

TABLE 8.7   MHS Basic (Base) Elements of Service

Service element	MT	MH/PD	MS	IPM
Access management	X	–	–	X
Basic physical rendition	–	X	–	–
Content type indication	X	–	–	X
Converted indication	X	–	–	X
Delivery time stamp indication	X	–	–	X
IP-message identification	–	–	–	X
Message delivery	X	–	–	–
Message identification	–	–	–	X
Nondelivery notification	X	–	–	X
Ordinary mail	–	X	–	–
Original encoded information types	X	–	–	X
Physical forwarding allowed	–	X	–	–
Stored message deletion	–	–	X	–
Stored message fetching	–	–	X	–
Stored message listing	–	–	X	–
Stored message summary	–	–	X	–
Submission time stamp indication	X	–	–	X
Typed body	–	–	–	X
Undeliverable mail with return	–	X	–	–
User/UA capabilities registration	X	–	–	X

The MHS 1984 and 1988 Recommendations

Even though the 1984 MHS version had some parts that were for "further study," it was a reasonably complete messaging protocol. Nonetheless, the 1988 release represents a substantial change to MHS, principally in the following areas.

The principal changes in the 1988 standard includes additional service elements for MT and IPM. From the standpoint of adding new features, this is probably the biggest change. The MHS changes also reflect the decision of ITU-T to better align these recommendations with the OSI architecture, especially the architecture of the upper three layers of OSI. It also was decided that some of the protocols in MHS were performing services that were applicable to systems other than MHS. These protocols were removed from MHS and placed into the X.200 Recommendations.

In addition, these MHS functions have been changed:

- A considerable number of security services added
- A message store service added
- A physical delivery service added
- The addition of distribution lists
- The reliance on the use of a directory
- The addition of redirection facilities
- Removal of the ASN.1 and Basic Encoding Rules (BER)

MHS and Electronic Data Interchange (EDI): X.435

The ITU-T approved X.435 in 1990. It defines a message handling application called Electronic Data Interchange messaging (EDIMG). The EDI standard is a very widely used specification. It is published under ANSIX12 and ISO's EDIFACT specification (ISO 9735). It provides rules for the creation, sending, and reception of standardized messages dealing with the business environment. For example, purchase orders, customs forms, commercial invoices, insurance forms, banking forms are candidates for EDI. X.435 defines the EDI messaging objects (in ASN.1), as well as the ports, and errors associated with EDI operations. The reader must have some knowledge of ASN.1 if this Recommendation is to be understood.

X.435 defines a protocol called PEDI, which defines how X.400 supports the exchange of EDI documents (see Figure 8.13). The essential aspect of PEDI is that it defines a new message content type specific to EDI. In addition, the ITU-T has also published an F Series document, F.435 "MHS EDI Message Service."

The 1992 MHS release also defines five security elements pertaining directly to EDI message exchange. The *proof of EDI notification* is used by

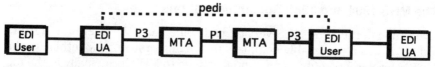

Figure 8.13 PEDI and MHS.

the recipient of the EDI message to create a notification to authenticate the recipient. The (a) *nonrepudiation of the EDI notification* allows the recipient of the EDI message a proof of origin of the document. This element prevents the originator from denying that it has sent the document. The *proof of content received* allows the originator to authenticate that the message is indeed received and it was the same message that was sent. The (b) *nonrepudiation of content originated* is used to protect the recipient from an originator denying the document was sent (repudiation of the old story "I never sent it to you"). The non-repudiation of content received protects against the originator from the recipient denying that the traffic was received (the old problem of "I never received your message").

X.440: Voice Messaging System

The reader may wonder why voice messaging is included in MHS, because to this point, MHS has been described in terms of supporting electronic documents of some type, such as E-mail and business forms. However, nothing precludes the conveyance of electronic voice images through an MHS as well. To this end, X.440 defines the procedures and message formats to support this type of application.

Like the other parts of the X.400 Series, X.440 is organized around protocol ports. The voice messages are placed into P1 enveloped, and contain a heading, and body parts. Indeed, almost all of X.440 is devoted to defining the syntax (in ASN.1) of the protocol data units.

X.440 provides no guidance or information on the how the voice images are represented in the voice message. This operation is defined in G.721 (sampling, quantizing, encoding, etc.). Furthermore, it does not define how the voice messages are managed in regard to delay and playout at the receiving entity. These important aspects are independent of X.440, and must be handled in the user application.

Summary

The ITU-T MHS Recommendations are some of the most widely used recommendations standards in the world. The architecture of MHS is based on the OSI Model and MHS uses ROSE, RTSE, and ACSE. It also relies on the presentation and session layers of the OSI Model for numerous support features. Like many ITU-T protocols, MHS is organized around abstract services called ports. Ports are defined through the P3, P1, P7, and P2 protocols. The MHS service elements also represent the foundations of the standards. Each of these service elements was described briefly in this chapter.

MHS also is organized around MDs that facilitate the naming, addressing, and routing of messages in the MHS. Additionally, it makes extensive use of the X.500 Directory for many security features.

9

The X.500 Series

Directories have been in use in computer installations for over a decade. Some organizations have used them for simple operations such as storing source code for programs. Others have built data directories to store the names and attributes of the organization's data elements.

Introduction

Some forward-thinking companies now use directories to show the relationships of data elements to databases, files, and programs. The Directory (also called a dictionary by some vendors) is used to check all key automated systems for accuracy and duplication and permits an organization to access the impact of system changes on all the automated resources. The Directory has become a vital component in an organization's management of its automated resources.

Typically, each organization and vendor has developed a unique and proprietary approach to the design and implementation of directories, which greatly complicates the management of resources that are stored in different machines and databases. In the spirit of OSI, the purpose of the X.500 Directory is to provide a set of standards to govern the use of directories and dictionaries.

The OSI Directory (X.500)

The X.500 Recommendations describe the operations of the Directory. It is designed to support and facilitate the communication of information between systems about *objects* such as data, applications, hardware, people, files, distribution lists, and practically anything else that the organization deems worthy of "tracking" for management purposes. X.500 is intended to allow the communication of this information between different systems, which can include OSI applications, OSI layer entities, OSI management entities, and communications networks.

X.500 encompasses eleven recommendations, collectively known as the X.500 Recommendations. Their full titles are listed below. Their general organization is shown in Figure 9.1.

As with some of the other X Series, the X.500 Series individual specifications are closely correlated with each other. Therefore, this chapter examines them as a logical whole, and not on a one-to-one basis.

Directory System Protocol—Protocol Implementation Conformance Statement (PICS) Proforma: X.582

This specification was released in the fall of 1992. As its name suggests, it defines the procedures for an organization (such as a test lab) to verify that a supplier's product conforms to the 1988 X.500 Recommendations. It also explains which other protocols may be affected by an application context in which X.500 is used, and provides reference to several sections in ROSE (X.249) and ACSE (referencing only ISO/IEC DIS 8650-2, because the ITU-T does not yet publish a PICS for ACSE).

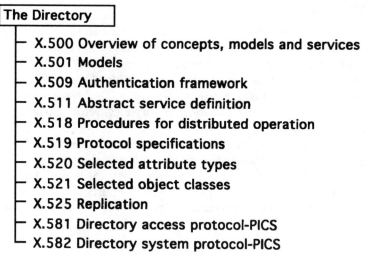

The Directory

- **X.500 Overview of concepts, models and services**
- **X.501 Models**
- **X.509 Authentication framework**
- **X.511 Abstract service definition**
- **X.518 Procedures for distributed operation**
- **X.519 Protocol specifications**
- **X.520 Selected attribute types**
- **X.521 Selected object classes**
- **X.525 Replication**
- **X.581 Directory access protocol-PICS**
- **X.582 Directory system protocol-PICS**

Figure 9.1 The organization of the X.500 Series.

- X.500: The Directory—Overview of concepts, models, and services

- X.501: The Directory—Models

- X.509: The Directory—Authentication framework

- X.511: The Directory—Abstract service definition

- X.518: The Directory—Procedures for distributed operation

- X.519: The Directory—Protocol specifications

- X.520: The Directory—Selected attribute types

- X.521: The Directory—Selected object classes

- X.525: The Directory—Replication

- X.581: Directory access protocol—Protocol implementation conformance statement (PICS)

- X.582: Directory system protocol—Protocol implementation conformance statement (PICS)

X.500 Services to Manage the Computer Communications Environment

As just discussed, the X.500 Recommendation places no requirement on the nature of the information stored in the Directory. However, organizations most likely will use it to provide a wide variety of services. Table 9.2 contains a list of likely applications for the Directory. This list reflects the author's ideas on the subject and also includes management services described in the ITU-T management standards (the X.700 Series).

TABLE 9.1 ITU-T/ISO Specifications for Directories

ITU-T	ISO
X.500	9594/1
X.501	9594/2
X.509	9594/8
X.511	9594/3
X.518	9594/4
X.519	9594/5
X.520	9594/6
X.521	9594/7

TABLE 9.2   X.500 Directory Support Services (Likely Applications, not All-Inclusive)

Service	Description
Configuration data	Description of network components and their status. Information on active, passive, and best paths in the internet.
Attributes	Characteristics of network resources
Names and aliases	Storage of user-friendly names as well as the mapping to network specific names; ability to cross-reference other names in other networks
ASN.1 code	ASN.1 coding for protocol data units (PDUs) and macros
Alarm storage	Rules for network alarm measurement, severity analysis, and alarm clearing
Fault data	Storage of data on faults, and error logs
Accounting data	Information on usage, charges of network components
Security features	Rules for authentication, keys for encipherment, security logs
Performance	Rules for performance analysis, as well as statistics on network behavior

Key terms and concepts

Our task in this section of the chapter is to come to grips with several key terms, concepts, and definitions used by the X.500 specifications (see Figure 9.2). The information held in the Directory is known as the *Directory Information Base* (DIB). The DIB contains information about objects and is composed of entries. Each *entry* consists of a collection of information on one object only. Each entry in made up of *attributes,* and each attribute has a *type* and one or more *values.*

The DIB is accessed by the Directory user through the *Directory User Agent* (DUA), which is considered to be an applications process. The DUA is so named because it acts as an agent to the end user *vis-á-vis* the DIB.

Figure 9.3 shows that the entries in the DIB are arranged in a tree structure called the *Directory Information Tree* (DIT). The vertices in the tree represent the *entries* in the DIB. These entries make up a collection of information about one *object* (such as a person, a data element, a piece of

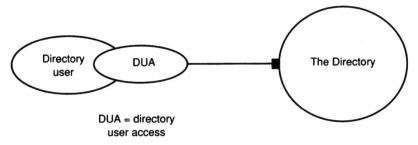

DUA = directory
user access

Figure 9.2 Functional view of the X.500 Directory.

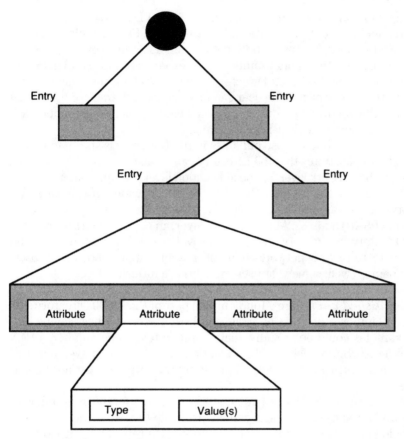

Figure 9.3 The Directory information tree.

hardware, a program, and so on). The alias is also permitted (not shown in the tree); it points to an object entry.

As we just learned, each entry in the DIT consists of *attributes,* and each attribute is made up of a *type* and one or more *values.* The attribute type specifies the syntax and the data type of the value (such as integer, Boolean, etc.).

Directory names. A fundamental concept of X.500 is user-friendly naming. Names are stored in the Directory, and the Directory provides the service of verifying names and uses a name to retrieve addresses. These addresses could be any type of address, such as an ISDN address, an X.25 address, an X.121 address, etc. The Directory uses the name for verification, retrieval of addresses, searching and browsing, and so forth.

Perhaps a good analogy to the use of the X.500 Directory can be made with the telephone system's white and yellow pages. For example, when using the white pages, a user-friendly name such as John Brown can be accessed to obtain a telephone number and perhaps an address. This same idea holds true for the X.500 Directory services. Taking the analogy a bit further to the yellow pages, the user can browse through trade or functional names and the yellow page "directory" will yield the telephone number, address, and of course the user-friendly name.

A typical approach to the use of names in the Directory is shown in Figure 9.4. This figure illustrates that the Directory can be accessed at either the application or the network layer. At the application layer, the application process provides a name to the Directory, most likely an application title. In turn, the Directory uses the application title name to retrieve a presentation service access point (PSAP) address. At the network layer, an NSAP address is provided to the Directory by the network service provider, which is then used by the Directory to retrieve a network point of attachment. In both cases, user-friendly names (well, somewhat user-friendly) are furnished to the Directory, and the Directory returns PSAPs and network points of attachment.

Distinguished names are a fundamental part of the X.500 Directory service. X.500 imposes no rules on what the distinguished names may be, but some examples could be an individual's name, a telephone number, a FAX number, a distribution list, an OSI network address, etc. However, a distinguished name must uniquely identify a Directory entry. It must not be ambiguous.

X.500 uses the term *schema* in a context that is different from other directory and database systems. A schema is a rule to ensure the DIB maintains its logical structure during modifications. It prevents inconsistencies in the DIB such as incorrect subordinate entries' class, attribute values, etc.

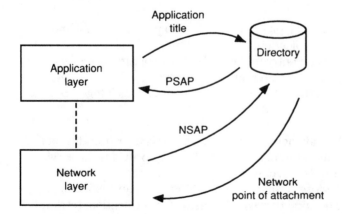

Figure 9.4 X.500 mapping of names.

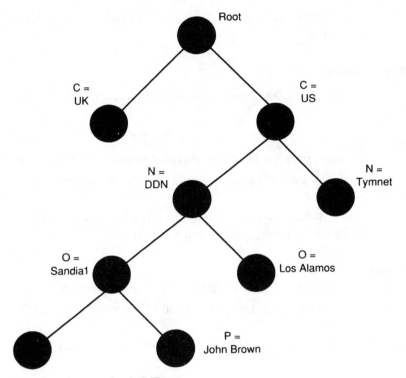

Figure 9.5   An example of a DIT.

It is the responsibility of the Directory to ensure that any changes to the DIB are in conformance with the Directory schema. Furthermore, controls are present to prevent a user from exceeding thresholds such as the scope of a search, the time spent on a search, the size of the results, etc.

The DIB may return a request using more than one entry. To do so, X.500 uses a *filter*. Many network control systems use the concept of filters. A filter is an assertion or a set of assertions about the attributes of a resource (in OSI terms, a managed object). In its simplest terms, it is a way to select tests for the selection of operations. The filter operations can be coded into C, ASN.1, etc., to select operating parameters from the Directory. Filters use the conventional Boolean operators of AND, NOT, and OR.

Figure 9.5 shows an example of a DIT with type attributes to identify the objects in the tree. The name C = US, N = DDN, O = SANDIA1, P = John Brown identifies the entity John Brown. Each node is identified by a relative distinguished name (RDN).

The entity is located at Sandia1, accessed through the Defense Data Network (DDN), within the US. The entity has as its naming attributes, the naming attributes of the next higher vertix in the tree: Sandia1. Thus, the distinguished name is US.DDN.Sandia1.John Brown, which is a concatenation of the node names.

Directory Services

At the broadest level, the X.500 Recommendation defines the following operations:

Service quality

- Controls: Establishes the rules for access and modification to the Directory
- Security: Defines the authentication, password, and other security procedures for Directory use
- Filters: Rules that define one or more conditions that must be satisfied by the entry if it is to be returned in the result to the user

Interrogation

- Read: Obtains the values of some or all the attributes of an entry
- Compare: Checks a value against an attribute of a specific entry in the DIB
- List: Obtains and returns a list of subordinates of a specific entry in the DIB
- Search: Using a filter, a certain part of the DIT is returned to the user
- Abandon: Causes the Directory to cease processing a prior request from a DIB user

Modification

- Add entry: Adds a new leaf entry to the DIT
- Remove entry: Removes a leaf entry from the DIT
- Modify entry: Changes a specific entry, including the addition, deletion for replacement of attribute types, and attribute values
- Modify relative distinguished name: Changes a distinguished name of a leaf entry in the DIT
- Other errors: Rules that describe how and why errors are reported
- Referrals: Procedures for referencing other resources

X.500 Services and Ports

Like other ITU-T Recommendations, the Directory is organized around the concept of ports. As discussed in earlier chapters, a port represents the service seen by the user of an ITU-T protocol.

The X.500 ports are organized as shown in Figure 9.6. Three ports are defined: read, search, and modify. The operations for each port are shown in Figure 9.7.

The *read* port consists of three operations: (1) read, (2) compare, and (3) abandon.

The purpose of the *read* operation is to extract information from a Directory entity. The operation must contain the identity of the entry from which the information is to be extracted. It must also contain an argument

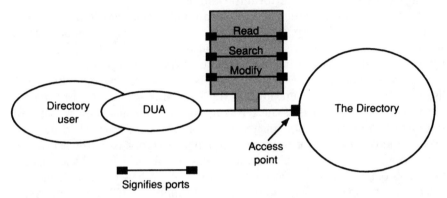

Figure 9.6   The X.500 ports.

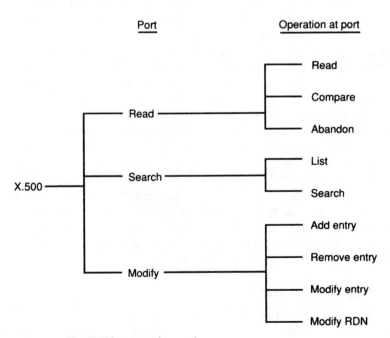

Figure 9.7   The X.500 ports and operation.

that identifies what information is to be extracted from the named object. The result of this operation contains an entry parameter that holds the information that was obtained as a result of the request. There are also several fields in this operation for listing status indicators and error reports in the event of problems.

The *compare* operation is used to compare an argument value in the request of the operation in an attribute type in the Directory. As with all operations to the Directory, the name of the object must be provided, as must be the arguments used to perform the compare.

The last operation for the Directory read port is the *abandon* operation. Its purpose is self-descriptive: It is used to abandon a previous operation. Although listed with the read port, it is also applicable to the search port (specifically, the list and search operations of the search port).

The *search* port, as indicated in Figure 9.6, consists of two operations: (1) list and (2) search. As its name implies, the list operation obtains a list from the Directory. It is used to access the Directory for the immediate subordinates of the identified entry. A distinguished name or relative distinguished name is used to identify what set is to be accessed in a Directory. Also, the list operation contains time and size limits which, if exceeded, will cause the list operation to stop. If the list is not finished, partial results will be made available.

The *search* operation is used to browse through the DIT and to return information from the entries. This operation identifies the node in the tree where the search begins. It could possibly be the root of the DIT. In addition, the search operation allows the user to define the searches to be applied to subordinates above or below a node.

The *modify* port consists of the following operations: (1) add entry, (2) remove entry, (3) modify entry, and (4) modify RDN. *Add entry* allows the addition of a leaf entry to the DIT. The arguments for this operation define the entry that is to be added and provide information about where the entry should reside in the Directory tree.

The *remove entry* simply performs the reverse operation of the add entry; it removes a leaf entry from the DIT.

The *modify entry* operates on an existing entry in the DIT. It is used for a number of modifications that could be made to the Directory: (1) removing an attribute, (2) adding an attribute value, (3) adding a new attribute to an existing entry, (4) removing an attribute value, (5) modifying an alias in the Directory, and (6) replacing attribute values.

The *modify RDN* is used to change an RDN of a leaf entry. It may be either an object entry or an alias entry in the Directory.

X.500 Error Reporting Services

The Directory defines a number of error-reporting procedures. It is quite important that they be implemented as part of the ongoing operations because they provide important diagnostic and troubleshooting services.

The Directory ceases to perform operations once an error has occurred and has been detected. It is possible that more than one error may be detected. In this case, the error is reported in accordance with a logical precedence list.

The higher entry in the following list means the error reporting has precedence over the next lower entry. Following these lists, each error is discussed.

- NameError
- UpdateError
- AttributeError
- SecurityError
- ServiceError

The following errors do not use any reporting precedences:

- AbandonFailed
- Abandoned
- Referral

The *NameError* is used to report any error dealing with a name. This operation is invoked if any one of six conditions are detected:

1. A name supplied in an argument to access the Directory does not match the name of any object in the Directory.

2. An alias problem has been encountered (the alias has no object).

3. An attribute type is not defined.

4. An attribute type and an accompanying attribute value are not compatible.

5. An invalid list operation has been encountered.

6. An alias problem has been encountered (the alias is not allowed).

The *UpdateError* is used to report any error relating to the operations of add, modify, or delete. This operation is invoked if any one of seven conditions are reported:

1. An attempted modification or addition to the Directory is not in accordance with the rules of the DIT.

2. An attempted update is not consistent with the definition provided by the object class.

3. An attempted operation is not permitted (permitted only on leaf entries).

4. An attempted operation would adversely affect the RDN.

5. An attempted operation would create a redundant alias.

6. An attempted operation would affect multiple Directory Service Agents (DSAs) (not allowed).

7. An object class attribute cannot be modified.

An *AttributeError* is used to report a problem related to an attribute. The operation identifies the attribute type and the value that caused the problem. Five conditions are reported:

1. The named entry does not have one of the attributes specified in the argument.

2. An attribute value in the argument does not conform to the syntax of the attribute type.

3. An argument has an undefined attribute type.

4. A matching rule was used that is not defined for the attribute type.

5. An attribute value violates an attribute definition.

The *SecurityError* is used to report an error relating to a security problem. Six conditions are reported:

1. The requestor's credentials are insufficient for the required protection.

2. The requestor does not have the proper access rights.

3. The requestor's credentials are invalid.

4. The requestor's signature is invalid.

5. The argument is not signed.

6. An invalid service is used in the DSA service.

The *ServiceError* is used to report a problem about a service provision. The following problems are reported:

1. The Directory is not available.

2. The Directory is too busy to provide a requested service.

3. A chain operation is required to provide the service, but a chaining prohibited option is in effect.

4. The service is denied because it would consume excessive resources.

The *AbandonFailed* is used if an abandon operation is in effect and a problem is encountered. The following problems are reported:

1. The Directory has no knowledge about the operation that is requested to be abandoned.

2. The Directory has already responded to the operation.

3. The Directory is not allowed to abandon an operation or the abandon cannot be performed.

The *abandoned* is used to report any outstanding Directory inquiry operation if the DUA invokes it properly. The *referral* is not an error operation but is used to inform the DUA that it must present its request to another access point.

Operating a Distributed Directory

X.500 ensures that the Directory can be distributed across a wide geographical area. To support this environment, the DSA provides access to the DIB from the DUAs or other DSAs. Figure 9.8 shows the relationship in the distributed Directory. A DUA is permitted to interact with a single or multiple DSAs. In turn, the DSAs may internetwork with other DSAs through referrals to satisfy a request. The DSA is considered to be an OSI application process.

The DUA and DSA communications are governed by two protocols:

- Directory Access Protocol (DAP): Specifies actions between DUA and DSA in order for the DUA to have access to the Directory.

- Directory System Protocol (DSP): Specifies actions between DSAs.

The Directory is administered by the *Directory Management Domain* (DMD), which consists of a set of one or more DSAs and zero or more DUAs. The DMD may be a country, a PTT, a network, or anything designated to manage the DIB.

The distributed directory model uses the read, search, and modify ports of the Directory and also extends them to three other ports, called *chained service ports*. They are as follows: *chained read, chained search,* and *chained modify.* The purpose of the chained operations is to pass a request to another DSA because the passing DSA knows that the other DSA has knowledge about the operation.

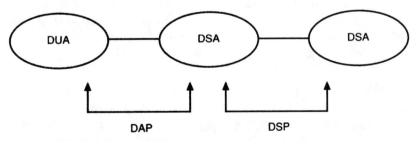

Figure 9.8 Distributed operations.

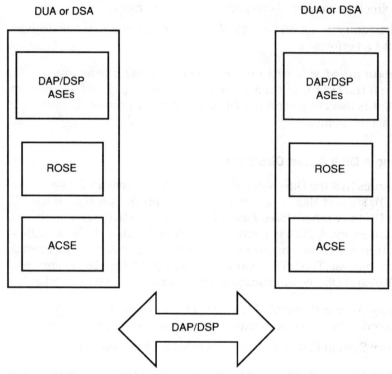

Figure 9.9 Directory protocol model.

The Relationship of the Directory to OSI (X.519)

The Directory can be used at two layers of the OSI Model. At the application layer, it supports the mapping of application-titles into PSAP addresses. At the network layer, it supports the mapping of NSAP addresses into subnetwork points of attachment (SNPA).

The Directory stipulates several directory-specific application service elements (ASEs) to support the DAP and DSP operations. These ASEs are involved in reading, comparing, searching, and the other Directory operations. In addition, DSP and DAP make use of the Remote Operations Service Element (ROSE), the Association Control Service Element (ACSE), and the OSI lower layers. The relationships are shown in Figure 9.9.

Use of ROSE

The Directory ASEs are the sole users of the following ROSE services: RO-INVOKE, RO-RESULT, RO-ERROR, RO-REJECT-U, and RO-REJECT-P. The operations of the DAP and DSP are class 2 operations. A ROSE category 2 is asynchronous and reports success (result) or failure (error). A DUA may choose to operate in class 1 synchronous and report success (result) or failure (error).

Use of ACSE

The Directory uses the ACSE A-ASSOCIATE and A-RELEASE services of the ACSE in normal mode. The application process uses the ACSE A-ABORT and A-P-ABORT services.

Use of the presentation layer

The Directory relies on ROSE and ACSE to obtain the services of the presentation layer. ROSE is the sole user of the P-DATA service. ACSE is the sole user of the P-CONNECT, P-RELEASE, P-U-ABORT, and P-P-ABORT services.

Use of other layers

The presentation layer uses the kernel and duplex functional units of the session layer. The use of the transport layer classes is optional, with one exception: ITU-T stipulates that (at a minimum) transport class 0 must be used. The use of X.213 for obtaining the network service is assumed. It is also assumed that a network address is used as stipulated in X.121, E.163-E.164, or X.200 (for the OSI NSAP address).

Authentication Procedures for Directory Access (X.509)

It is reasonable to expect an information repository as important as an information resource directory to have security features to prevent unauthorized access. In turn, a secure Directory might contain the authentication names and passwords to support other applications. The Directory includes these types of support features in X.509. Presently, two types of authentication are defined. *Simple* authentication uses a simple password authentication scheme, and *strong* authentication uses public key cryptographic techniques. The user may choose between simple and strong authentication, depending upon the need for secure services.

Simple authentication

Simple authentication is supported only within a simple directory domain and is restricted to use either one DUA and one DSA or two DSAs. The procedure for simple authentication operates as follows (see Figure 9.10). User A sends to user B its distinguished name and password (step 1). This information is then forwarded to the Directory and the password is checked against an appropriate password in the Directory (step 2). The Directory

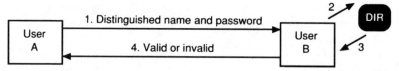

Figure 9.10 Simple authentication.

then informs B that A's credentials are valid or invalid (step 3). Finally, B informs A about the results of the Directory authentication operation (step 4).

Other procedures are also supported. For example, it is possible for B to check the distinguished name and password. Another approach is to use a random number or a time stamp with the distinguished name and password.

Strong authentication

The ITU-T has adapted the public key crypto system (PKCS) for the Directory strong authentication operations. The public key concepts were developed in the early 1980s and are now widely used throughout the industry. A brief tutorial follows on the public key concept.

Public keys have been in existence for just a few years. The advantage that public keys have over private keys is the ease of their administration and the ease of changing their values.

The concept of public keys is illustrated in Figure 9.11. As noted, the user A has, in its possession, the public key for user B's and A's private key. In turn, user B has in its possession the private key for user B and the public key for user A. It is essential that both A's and B's private and public keys be derived from the same function.

As shown in the figure, A can use B's public key to encrypt the data to be transmitted to B. In turn, B uses its private key (which is not available to any other user) to decipher the traffic. The reverse process occurs by B using A's public key for encryption before transmitting the data to A, which then applies its private key for decrypting the traffic into clear text.

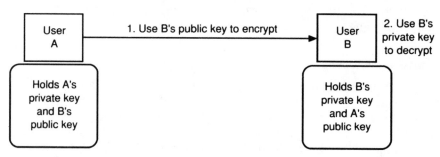

Figure 9.11 Public keys.

Digital signatures. This previous example shows the use of the public key for enciphering data and a complementary secret key used for deciphering the data. This process can be reversed to provide a very powerful authentication procedure, known as a *digital signature*.

A digital signature operation is illustrated in Figure 9.12. User A employs its private key to encipher a digital signature (for example, a password or some other form of identification). This is transmitted to user B, which has possession of user A's public key. Any user that has possession of the public key can decipher the data but only A can perform the complementary encryption because it alone has the private key.

User A first informs B that it is indeed A. Next, it sends the encrypted traffic. Then, B uses A's public key to decipher the protocol data unit. If A stated it was someone else, it would not have the proper private key to create the digital signature. In this manner, a user can be authenticated and the source of the information can be verified. Of course, because some of these concepts are new, most systems do not employ authentication and digital signatures. Notwithstanding, they are very powerful and flexible techniques and the OSI Model and the X500 Directory employ digital signatures for authentication.

X.509 requires that both the private and public keys be used for encryption and decryption in the following manner:

- Public key used to encipher; private key used to decipher
- Private key used to encipher; public key used to decipher

A specific encryption/decryption algorithm is not defined by the Directory. As long as the users involved in the authentication process use the same system, and each possesses a unique distinguished name, the Directory will support the authentication process. X.31 in chapter 4 contains more information on public encryption keys.

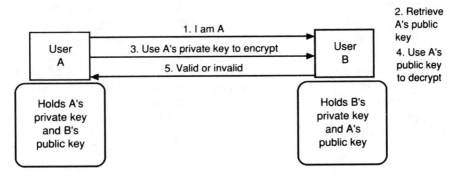

Figure 9.12 Digital signatures.

Example of the ASN.1 Code for a Port

A considerable portion of the Directory is devoted to abstract service definitions; X.511 is devoted exclusively to ASN.1. It describes the operations between the Directory and a DUA. As discussed earlier, these operations (actions) are defined with ports. An ASN.1 example of a port read operation is included in this section. The reader should refer to the X.500 documents for a complete description of all these operations.

A read extracts information from the DIB or verifies a distinguished name. It takes the following ASN.1 form:

```
Read :: = ABSTRACT-OPERATION
     ARGUMENT        ReadArgument
     RESULT      Read Result
     ERRORS  {
          AttributeError, NameError,
          ServiceError, Referral, Abandoned,
          SecurityError }
ReadArgument     :: =  OPTIONALLY-SIGNED SET {
          object          [0]    Name,
          selection       [1]
       EntryInformationSelection
                                    DEFAULT {}
               COMPONENTS OF CommonArgument }
ReadResult          :: =    OPTIONALLY-SIGNED SET {
          entry           [0]    EntryInformation,
          COMPONENTS OF     CommonResults }
```

The *ERRORS* have been described in a previous section of this chapter. The other parameters for this operation are described in the following paragraphs.

The *ReadArgument* contains the object name and selection criteria. The *object* argument identifies the object entry that is to be accessed for the requested information. The *selection* entry is a key search beyond the object argument, in that it further refines the search to define the requested information from the entry. This entry specifies the set of attributes and, optionally, information about the selected attribute will be returned. The *CommonArguments* are used to define (1) what controls are to be used during the operation (chaining, use of alias, etc.), (2) what security parameters are to be used (certification required, etc.), and (3) what role the DSA is to play in the operation. Finally, the *ReadResult* is used to report on the outcome of the operation.

Summary

The X.500 Recommendations define the use of a Directory to support ITU-T-based systems. The Directory is considered the linchpin for other systems, such as OSI network management and X.400. Like many of the more recently published ITU-T Recommendations, the X.500 Directory relies on the use of the application layer protocols, ROSE, and ACSE, as well as the presentation and session layers. Also, as with other ITU-T Recommendations, the services of the OSI Directory are defined through ports. The X.500 Directory also contains extensive documentation on authentication and security procedures for access to the Directory. It provides for simple authentication and strong authentication with public encryption keys.

10

The X.600 Series: OSI, Networking, and System Aspects

The X.600 Recommendations are relative newcomers to the ITU-T standards. X.612 was published in 1990, and the others have been added since that time. All of these specifications deal with a variety of issues pertaining to X.25, OSI, ISDN, naming, and name registration procedures. The organization of the X.600 Recommendations is shown in Figure 10.1, and their full titles are listed following.

Introduction

The reader might wish to review X.25 and X.31 in chapter 4 and X.223 in chapter 6 before reading about X.612. Figures 4.24 and 4.25 should be particularly helpful. This chapter assumes the reader is familiar with these recommendations.

- X.610: Provision and support of the OSI connection-mode network service

- X.612: Information technology—Provision of the OSI connection-mode network service by packet-mode terminal equipment connected to an integrated services digital network (ISDN)

- X.613: Information technology—Use of X.25 packet layer protocol in conjunction with X.21/X.21 bis to provide the OSI connection-mode network service

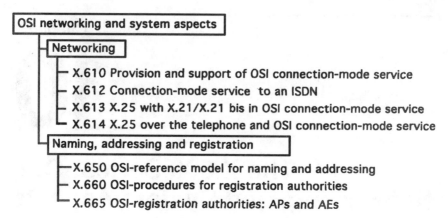

Figure 10.1   Organization of the X.600 Recommendations.

- X.614: Information technology—Use of X.25 packet layer protocol to provide the OSI connection network service over the telephone network

- X.650: Open Systems Interconnection (OSI)—Reference model for naming and addressing

- X.660: Information technology—Open Systems Interconnection-Procedures for the operation of OSI Registration Authorities—General procedures

- X.665: Information technology—Open Systems Interconnection-Procedures for the operation of OSI Registration Authorities: Application processes and application entities

X.610

X.610 is a general specification that provides an overview of the OSI X.213 connection-mode network service. Its purpose is to clarify (a) the relationship of the network layer service to the transport layer, and (b) the relationships of X.25, X.75, X.21, and X.21 bis in the connection-mode service. It is useful as a general model, and as an introduction to the other X.600 Series Recommendations. It explains the topology layouts and the associated layers for the configurations listed in Table 10.1.

X.612

X.612 defines the connection-oriented network service primitives and parameters that are passed between an OSI layer 4 and ISDN Q.931 or X.25 at layer 3.

X.216 assumes the use of X.31, both case A and case B. It also assumes the availability of semipermanent and demand B channels. Given these as-

TABLE 10.1 X.610 Configuration Options

Configuration
CSDN leased line access to a PSDN
CSDN switched line access to a PSDN
CSDN leased line
CSDN switched line connection
PSTN leased line access to a PSDN
PSTN switched line access to a PSDN
PSTN leased line connection
PSTN switched line connection
ISDN D channel access to a PH
ISDN B channel semi-permanent access to PH
ISDN B channel demand access to PH
ISDN B channel semi-permanent access to PSDN
ISDN B channel demand access to PSDN
ISDN B channel semi-permanent connection
ISDN B channel demand connection

sumptions, X.612 clarifies the use of X.223 between the third and fourth OSI layer and the permissible X.25 or ISDN stacks at the lower three layers. X.216 also describes the stack for using X.223 between OSI's fourth layer and the bottom three layers for an R reference point.

The general scheme of X.216 is illustrated in Figure 10.2. Figure 10.2*a* shows an X.31 case B access. The physical layer is I.430 for a basic access (2B+D) and I.431 for a primary access. Link access procedures for the D channel (LAPD) is used at the data link layer for D channel operations. At the network layer, X.931 may be invoked before the transmission of X.25 packets to set up resources and determine if an X.25 Incoming Call packet is to be conveyed on the B or D channel.

Figure 10.2*b* illustrates the option of using the D, B, or H channels. The physical layer signalling is with the basic rate or primary rate with I.430 and I.432, respectively. At the data link layer, Q.921 is used for D channel signalling and link access protocol, balanced (LAPB) is used on the B channel. In addition, this option uses Q.931 at the network layer to support circuit-switched access to the packet handler. This option supports both semipermanent B channel access (nonswitched service) and demand B channel access (switched service).

Figure 10.2*c* shows the protocol layers for X.25/ISDN interworking at the R reference point. This stack is the conventional X.25 protocol set. The TE2 is unaware if case A or case B access is being used because the terminal adapter (TA) provides the conversion functions. X.612 establishes the rules for three underlying physical connections for the R reference point: (1) a leased circuit, (2) a direct call, and (3) a circuit-switched call. The required X.21 states and the required V.24 circuits for X.21 bis and the V Series modems are specified in this section of the recommendation.

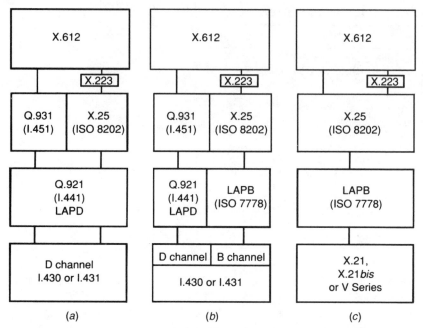

Figure 10.2 The X.612 protocol layers. (*a*) Using the D channel at S/T reference points; (*b*) using the B channel at S/T reference points; (*c*) protocols at the R reference points.

For the network designer, one of the most valuable aspects of X.612 is the mapping of the Q.931 cause codes to the X.213/223 reason fields in the primitive parameters. If X.21 is employed, it shows the mapping of the X.21 progress signals to these parameters. X.612 provides two tables to show both of these mapping relationships.

X.216 also defines the rules for mapping X.121 and E.164 addresses. The mapping procedures are in accordance with the X.300 Recommendations.

X.613

X.613 defines the conventions for using X.25 packet layer procedures (the X.25 layer three, and the OSI connection-mode network service) with X.21 or X.21 bis. The Recommendation is applicable to two environments: (a) a circuit switched data network (CSDN), or (b) a packet switched data network. These networks may be private or public. Be aware that X.613 assumes the use of the X.30-based ISDN terminal adapter (TA).

The X.613 protocol stack is similar to the stack depicted in Figure 10.2*c*. The one difference is that X.613 does not define a V Series at the physical layer—only X.21 or X.21 bis.

As with X.612, the network designer will find X.613 quite useful with its inclusion of the mapping of the X.21 call progress signal to the associated CONS packet parameter.

X.614

This recommendation defines the procedures for using X.25 layer three operations to obtain OSI CONS services over the telephone network. Four environments are defined: (a) CONS through a telephone network leased line, (b) CONS through a telephone switched (dial-up) line, (c) CONS to a PSDN through a telephone network leased line, and (d) CONS to a PSDN through a telephone switched line. For this last environment, it is assumed that X.32 is employed for the exchange of identification parameters. In addition, X.614 establishes the rules for using ISO/IEC 8208, which defines how a DTE plays the "role" of a DTE or a DCE.

One other point the reader should note about X.614 concerns the rules for addresses. For a direct DTE-DTE configuration through the telephone network, the X.25 address fields are not used. The called and calling NSAP values are contained in the called and calling address extension facilities. In addition, conventional telephone network addressing is used as the connection identifier between the two subnet points of attachment (SNPAs). In contrast, for a DTE-DCE configuration, the X.25 address field is used, as well as the called and calling address extension facilities, and telephone network addressing. The reader should obtain X.223 if more information is needed on these scenarios.

X.650

One of the complaints about several of the initial OSI documents dealing with naming and addressing was the vagueness of the specifications. Additionally, some of these recommendations were not complete. With X.650, the ITU-T has made "repairs" to its earlier Recommendations. In effect, it broadens the statements in X.200, and in some instances, replaces definitions contained in X.200. These changes are reflected in the revisions to X.200.

A discussion of each OSI address and name is beyond this general reference guide. The reader is encouraged to study X.650 for more information on the relationships of SAPs, names, selectors, and invocations. A brief description here will give you an idea of the information contained in X.650. I shall not repeat the information about this subject that is provided in chapter 2.

First, a selector is an element of addressing information that identifies a grouping (a set) of SAPs, that operate within one identifiable system (that is, one with an unambiguous address). The selector value is not registered, but is assigned at a system by a local administrator.

Second, an unambiguous address is called protocol addressing information or PAI. The PAI is constructed of the following tuple form: P-selector, S-selector, T-selector, and a network address (or addresses). The n-selectors represent values assigned to PSAPs, SSAPs, and TSAPs, where P = presentation layer, S = session layer, and T = transport layer. It can be seen that the PAI is a network address (or addresses) that allows the system to transport (or relay) the traffic to the DTE (the end user machine). The selectors are used to invoke the proper entities (software modules) in the respective upper layers.

Third, an entity (such as a software module) is identified by a name, which is called an entity title. A title directory is used to correlate this title to a PAI. This operation is shown in Figure 10.3.

X.650 also provides in-depth definitions and descriptions of naming and addressing for all seven layers of the OSI Model, with particular emphasis on the application layer names, titles, and invocation identifiers.

X.660

X.660 establishes the procedures for the registration of objects within an OSI environment. An object is anything that a network administrator wishes to be identified. As examples, it might be a piece of hardware, or a software module, or even a routing table in a switch. The unambiguous name associated with an object is called an object identifier. X.660 also defines the use of naming-domains, and how registration authorities operate with these domains.

This Recommendation also describes a naming tree that is used as a basis for a registration hierarchy (RH) of object identifiers (this scheme is also described in X.208). Figure 10.4 shows the top level hierarchy for the registration hierarchy naming tree. The top level arcs shown in this figure are assigned the integer values based on X.208. Once an organization (or some

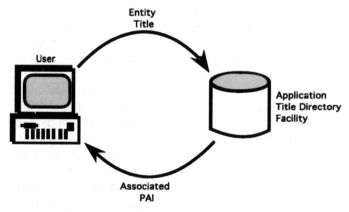

Figure 10.3 Relationship of Title and PAI.

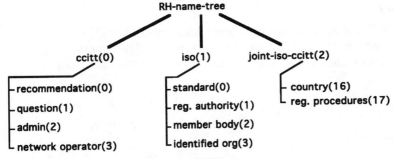

Figure 10.4 The Registration Hierarchy (RH) tree.

entity) has been registered under one of the nodes, the entity has been delegated that part of the tree, and can then administer any subtrees beneath it as it deems necessary.

The idea of the registration procedure is to provide a means for assigning a unique, unambiguous identifier to an object. For example, assume that a company in the US is to be identified, with the following name:

Country = US
State = Hawaii
Organization = Acme
Department = R&D

Using X.660, Annex C, as a guideline, the registration would appear as shown in Figure 10.5. Therefore, the full distinguished name would be a concatenation of the node values of: {2 16 840 46 2222 11}.

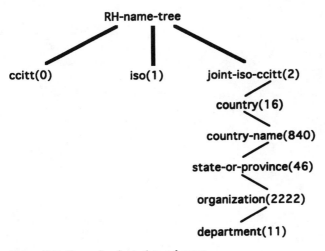

Figure 10.5 Example of a registered name.

Summary

The X.600 Recommendations are recent additions to the ITU-T Standards. They are quite useful to network administrators and designers due to their descriptions of address registrations and rules for names and addresses.

11

The X.700 Series OSI Management Standards

The International Standards Organization (ISO) has been working on the development of several Open Systems Interconnection (OSI) management standards for a number of years. Fortunately, the ITU-T has aligned its X.700 Series with the ISO Standards.

Introduction

This chapter introduces the ITU-T X.700 Recommendations and compares them to their ISO counterparts. It also discusses the conceptual foundations for these standards. A more thorough analysis of this subject can be found in another book in this series, *Network Management Standards* by Uyless Black.

The X.700 Series are founded on the OSI Model X.200 Recommendations. Indeed, the X.700 Recommendations are called the OSI management standards. Therefore, the reader should be familiar with the material in chapter 6 before reading this one.

Figure 11.1 shows the organization of the X.700 Series, and the list below shows the full titles.

- X.700: Management framework for Open Systems Interconnection (OSI) for CCITT applications

- X.701: Information technology—Open Systems Interconnection-Systems management overview

OSI management

- X.700 Management framework
- X.701 Systems management overview
- X.710 Common management information service (CMISE)
- X.711 Common management information protocol (CMIP)
- X.712 Protocol implementation conformance statement proforma
- X.720 SMI: Management information model
- X.721 SMI: Definition of management information
- X.722 SMI: Guidelines for definition of managed objects
- X.723 SMI: Generic management information
- X.724 SMI: Implementation conformance statement proformas
- X.730 SM: Object management function
- X.731 SM: State management function
- X.732 SM: Attributes for representing relationships
- X.733 SM: Alarm reporting function
- X.734 SM: Event report management function
- X.735 SM: Log control function
- X.736 SM: Security alarm reporting function
- X.738 SM: Summarization function
- X.739 SM: Metric objects and attributes
- X.740 SM: Security audit trail function
- X.745 SM: Test management function

Where:
SMI is Systems management overview
SM is Systems management

Figure 11.1   Organization of the X.700 Recommendations.

- X.710: Common management information service definition for CCITT applications

- X.711: Common management information protocol specification for CCITT applications

- X.712: Information technology—Open Systems Interconnection-Common management information protocol: Protocol implementation conformance statement proforma

- X.720: Information technology—Open Systems Interconnection-Structure of Management Information: Management information model

- X.721: Information technology—Open Systems Interconnection-Structure of Management Information: Definition of management information

- X.722: Information technology—Open Systems Interconnection-Structure of Management Information: Guidelines for the definition of managed objects

- X.723: Information technology—Open Systems Interconnection-Structure of Management Information: Generic management information

- X.724: Information technology—Open Systems Interconnection-Structure of Management Information-Requirements and guidelines for implementation conformance statement proformas associated with OSI management

- X.730: Information technology—Open Systems Interconnection-Systems Management: Object management function

- X.731: Information technology—Open Systems Interconnection-Systems Management: State management function

- X.732: Information technology—Open Systems Interconnection-Systems Management: Attributes for representing relationships

- X.733: Information technology—Open Systems Interconnection-Systems Management: Alarm reporting function

- X.734: Information technology—Open Systems Interconnection-Systems Management: Event report management function

- X.735: Information technology—Open Systems Interconnection-Systems Management: Log control function

- X.736: Information technology—Open Systems Interconnection-Systems Management: Security alarm reporting function

- X.738: Information technology—Open Systems Interconnection-Systems Management: Summarization function

- X.739: Information technology—Open Systems Interconnection-Systems Management: Metric objects and attributes

- X.740: Information technology—Open Systems Interconnection-Systems Management: Security audit trail function

- X.745: Information technology—Open Systems Interconnection-Systems Management: Test management function

Object-oriented design (OOD)

The OSI management standards use many of the concepts of OOD, which originated in the early 1970s. The notion of an object as a construct for manipulation (in effect a programming construct) was first found in Simula, which was a language used to program computer simulations. A significant event in the early history of OOD occurred in 1983, with the implementation of the Smalltalk language. Smalltalk operates on software objects. The idea of OOD (and Smalltalk) is that a program operates on an object without knowledge of the internal operations of that object.

This idea is shown in Figure 11.2. Two objects are communicating with each other and do not know about the internal operations of each other. They operate at the visible interface through the passing of messages (shown as operations). The internal realization (i.e., the operations within the objects) is hidden from other objects. The notion of OOD and operations on objects is found in OSI management's use of managed objects.

Managed Objects

The resources that are supervised and controlled by OSI management are called *managed objects*. A managed object can be anything deemed important by organizations that are using the OSI management standards. For example, hardware such as switches, work stations, PBXs, PBX port cards, and multiplexers can be identified as managed objects. Software, such as queuing programs, routing algorithms, and buffer management routines, can also be treated as managed objects.

Managed objects are classified by how they fit into the OSI layers. If they are specific to an individual layer, they are called *(N)-layer managed objects*. If they pertain to more than one layer, they are called *system managed objects*.

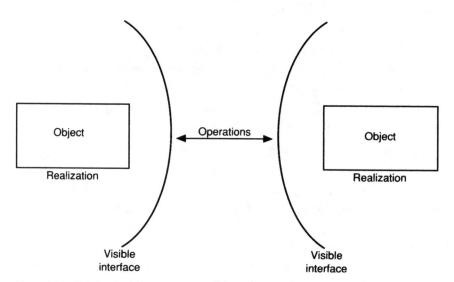

Figure 11.2 Object-oriented design concepts.

A managed object is completely described and defined by four aspects of OSI management:

- Its *attributes* (characteristics) that are known at its interface (visible boundary)
- The *operations* that may be performed on it
- The *notifications* (reports) it is allowed to make
- Its behavior exhibited in response to operations performed on it

Attributes

Managed objects have certain properties that distinguish them from each other. These properties are called *attributes*. The purpose of an attribute is to describe the operational characteristics, current state, and conditions of operation of the managed objects. Associated with the attributes are attribute values. For example, an object (such as a PBX line card) may have an attribute called status and a value of "operational."

Each attribute consists of one type and one or more values. For example, a type might be labeled the operational state of a packet switch. The values for this type could be disabled, enabled, active, or busy. As another example, the packet switch might have a type labeled the management state and the values for this attribute type could be locked, unlocked, or shutting down.

OSI management places a restriction on how attributes are manipulated. Attributes cannot be created, deleted, or named during the existence of an instance of a managed object (instances are described shortly).

Operations

OSI management establishes the permissible operations that can be performed on a managed object. The operations are:

- Create: Creates a new managed object
- Delete: Deletes a managed object
- Action: Performs an action (operation) on a managed object
- Get value: Obtains a value about an attribute of the managed object
- Add value: Adds a value to an attribute of the managed object
- Remove value: Removes a value from a set of values about a managed object
- Replace value: Replaces a value or values for an attribute of a managed object
- Set value: Sets default values for an attribute of a managed object

Notifications

Managed objects are permitted to send reports (notifications) about events that occur of which the object is aware. The nature of the notification depends on how the managed object is classified within the managed network.

Behavior

A managed object may also exhibit behavioral characteristics. These include (1) how the object reacts to operations performed on it and (2) constraints placed on its behavior.

The behavior of a managed object defines the conditions under which the values of the attributes may be altered by stimuli. Two kinds of stimuli can occur in relation to a managed object. First, external stimuli can occur through operations on the managed object. External stimuli consists of messages being passed to the visible boundary of the object. Second, the behavior of the managed object may be affected by internal stimuli. In this situation, the managed object does not have a message passed to it, but its behavior is affected by events such as internal timers.

The behavior of a managed object is defined also by the notifications it is allowed to create and send to some other entity. For example, it may not be allowed to report on a condition until the condition reaches or exceeds a certain threshold (such as an excessively busy CPU at a switch).

Management Process and Agent Process

Managed objects are managed by a *management process*. A management process is an application process (a network management program). As shown in Figure 11.3, the management process is categorized as either (1) a *managing process* or (2) an *agent process*. A managing process is defined as part of an application process that is responsible for management activities (a broad definition, to be sure). An agent process performs the management functions on the managed objects at the request of the managing process.

The communications between the managing process, the agent process, and the managed objects consist of (1) management operations and (2) notifications. It is not required for the agent process to know about the specifics of the management operations it receives or the notifications it

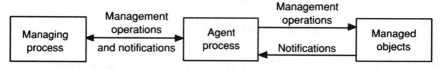

Figure 11.3 OSI management processes.

sends. For example, the agent process may send a notification with a value in it. The agent process need not know if the value is for an alarm or a performance parameter or whatever..

Object Classes

Managed objects that have similar characteristics are grouped into an *object class*. The characteristics used to determine an object class are attributes, operations, behavior, and notifications. Managed object classes provide a convenient means to group related resources together, and through the use of hierarchical naming techniques, it is possible to derive new classes from existing classes. This means that it is possible to encapsulate (or contain) objects within other objects and, in so doing, invoke operations or receive notifications only on the relevant "layer" of the encapsulated objects.

Layer Architecture of OSI Management

The OSI Model forms the underlying structure for the standards. For our purposes in this chapter, we will restrain the discussion to a general view of the subject.

An application-entity is involved in OSI management. It is called the *systems management application-entity* (SMAE). It is responsible for implementing the OSI System management activities. An SMAE is a collection of cooperating application service elements (ASEs).

The OSI management model is consistent with the overall OSI application layer architecture. One configuration is shown in Figure 11.4. Other configurations are permissible.

The systems management application service element (SMASE) creates and uses the protocol data units (PDUs) transferred between the management processes of the two machines. These data units are called management application protocol data units (MAPDUs) and are examined later in the book.

The SMASE may use the communications services of application services elements or the common management information service element (CMISE). As shown in the figure, the use of CMISE implies the use of ROSE and ACSE.

In accordance with OSI conventions, two management applications in two open systems exchange management information after they have established an application context. The application context uses a name that identifies the service elements needed to support the association. The application context for OSI management associations implies the use of Association Control Service Element (ACSE), Remote Operations Service Element (ROSE), CMISE, and one or more SMASEs.

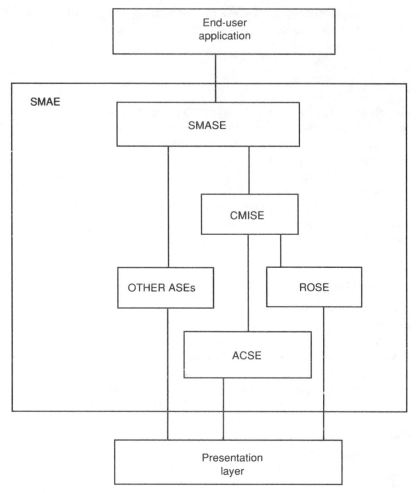

Figure 11.4 The OSI management structure.

Structure of the OSI Management Standards

Table 11.1 lists the OSI management standards. The ISO counterpart is also shown. This section provides an overview of each.

X.700 management framework

The foundation OSI management document is X.700. It provides the concepts and definitions for OSI management. It also introduces the five major functional components of OSI management: accounting, security, configuration, fault, and performance. In addition, it explains the concepts of indi-

TABLE 11.1   Relationship of the X.700 Series and the ISO Standards

Title	ISO	CCITT
Management Framework	7498-4	X.700
Systems Management Overview	10040	—
Structure of Management Information		
Part 1: Management Information Model	10165-1	X.720
Part 2: Definition of Management Information (10165-2)	10165-2	X.721
Part 4: Guidelines for Definition of Managed Objects (10165-4)	10165-4	X.722
Common Management Information Service Element (CMISE)	9595	X.710
Common Management Information Protocol (CMIP)	9596	X.711
Systems Management-Configuration Management		
Part 1: Object Management	10164-1	X.730
Part 2: State Management	10164-2	X.731
Part 3: Relationship Management	10164-3	X.732
Systems Management-Fault Management		
Part 4: Alarm Reporting	10164-4	X.733
Part 5: Event Reporting Management	10164-5	X.734
Part 6: Log Control Function	10164-6	X.735
Part __: Confidence and Diagnostic Test Classes	10164-_	X.737
Part __: Test Management	10164-_	
Systems Management-Security Management		
Part 7: Security Alarm Reporting Function	10164-7	X.736
Part 8: Security Audit Trail Function	10164-8	X.740
Part 9: Objects & Attributes for Access Control	10164-9	X.741
Systems Management-Accounting Management		
Part 10: Accounting Metering Function	10164-10	X.742
Systems Management-Performance Management		
Part 11: Workload Monitoring Function	10164-11	X.739
Part __: Measurement Summarization Function	10164-_	X.738
Part __: Software Management Function	10164-_	—
Part __: Time Management Function	10164-_	—

vidual layer management and the concepts of managed objects, which were introduced earlier in this chapter.

X.701 systems management overview

This standard identifies the underlying OSI services used by the management entities. It describes the concepts of distributed systems management, introducing the agent and management processes. X.701 also establishes the structure for the applications layer interactions among the ASEs.

X.701 defines the concepts of administrative management domains, which are similar to the administrative domains found in the ITU-T X.400 Recommendations. Real systems are organized into sets to: (1) meet certain requirement functions such as fault management, accounting management, etc., (2) assign the roles of the agent and manager, and (3) establish some kind of control over the process. They are called *functional management domains*. The functional management domains are under the control of a *management administrative domain*, which is responsible for the transfer of the control of resources between the functional management domains.

Application service elements and protocols

These standards are considered to be some of the more important parts of OSI management because they describe the service definitions (primitives) and the PDUs for the management operations. I mention them briefly here; ROSE and ACSE are covered in chapter 6; CMISE and CMIP are covered later in this chapter.

ROSE and ACSE. ROSE and ACSE are not part of the OSI management standards *per se*. Rather, they are standards published by the ISO and ITU-T for use as general software support systems for other applications. For example, the X.400 message handling system (MHS) standards use ROSE and ACSE.

X.710 Common Management Information Service Element (CMISE). CMISE is defined in X.710; it identifies the service elements (primitives) used in communications between the CMISE modules and their service users and providers. CMISE is described in more detail later in this chapter.

X.711 Common Management Information Protocol (CMIP). The X.711 Recommendation describes the protocol specification CMIP, which is the protocol counterpart to CMISE. It accepts the primitives from a network management application as defined by CMISE, constructs the appropriate APDU, and sends it (through the lower layers) to the peer CMIP user. Using the CMISE primitives, the receiving peer CMIP passes the APDU to the receiving network management application. CMIP is described in more detail later in this chapter.

OSI Management Functional Areas

For purposes of organization and documentation, five OSI management functional areas have been defined by the ITU-T and ISO. Be aware that these areas are actually contained in the separate documents listed in Table 11.1 and the reader should study these documents for more information on the functional areas. Notwithstanding, their organization provides a convenient means to summarize important aspects of these standards. The functional areas are:

- Fault management
- Accounting management
- Configuration management
- Security management
- Performance management

Fault management

Fault management is used to detect, isolate, and repair problems. It encompasses activities such as the ability to trace faults through the system, to carry out diagnostics, and to act upon the detection of errors in order to correct the faults. It is also concerned with the use and management of error logs. Fault management also defines how to trace errors through the log and time stamping of the fault management messages.

Fault management makes use of the OSI CMISE service element to provide for three primary activities:

- Fault detection: Faults can be detected by either monitoring or error report generation.
- Fault diagnosis: Faults are diagnosed by implementing diagnostics on a component (an OSI managed object) by (1) reproducing the error, (2) analyzing the error, or (3) receiving reports from the managed objects.
- Fault correction: Fault correction is accomplished through the use of other facilities, such as the configuration management facility.

Fault management relies on several other services (functions) defined in other OSI management standards (described shortly):

- Event reporting: Supports the transfer of event and error reports.
- Confidence and diagnostic testing: Supports the means to determine if a managed object is able to perform its function.
- Log control: A common function that supports activities such as managing event logs, restricting access to managed objects, and so forth.
- Alarm reporting: Supports the use of alarm reporting in the system.

Accounting management

This facility is needed in any type of shared resource environment. It defines how the usage, charges, and costs are to be identified in the OSI environment. It allows users and managers to place limits on usage and allows for the negotiation of additional resources. As of this writing, this functional area is not well defined, and it will probably be some time before the standard is approved as an international standard. However, the document contains enough detailed information for planning purposes.

Configuration management

This facility is used to identify and control managed objects. It defines the procedures for initializing, operating, and closing down the managed objects and the procedures for reconfiguring the managed objects. It is also used to associate names with managed objects and to set up parameters for the objects. Lastly, it collects data about the operations in the open system in order to recognize a change in the state of the system.

The configuration management standard defines the operational states of managed objects. Configuration management relies on X.731 for the definitions of states.

Four operational status states are defined (see Figure 11.5). If an organization uses the OSI network management standards, it is required to use these status indicators. The operational states are:

- Enabled: The resource (managed object) is not in use, but it is operable and available.

- Disabled: The resource is not available or it is dependent upon another source that is not available.

- Active: The resource is available for use and it has the capacity to accept services from another source.

- Busy: The resource is available but it does not have the spare capacity for other services.

The permissible operational states shown in Figure 11.5 are not found in all managed objects. Some objects have no limit on the number of simultaneous users that could use the object. In this situation, the managed object could not exhibit the busy state. As an example, in a connectionless gateway, an incoming queue for datagrams would never be considered busy because excess traffic is simply discarded. In addition, it is unlikely that a managed object would exhibit a disabled state if it has no dependencies on other managed objects because the state of "disabled" is irrelevant.

Certain rules are evident from an examination of Figure 11.5. For example, the transition to the enable state means the taking of actions that make the managed object operable. Actions, such as replacing a faulty line card, making a change to a software bug, etc., may permit the object to be declared operable. Notice that the transition to enable can only occur if the object's operational state is disabled. However, certain other activities allow an object to move to the enabled state. For example, a user may quit using a non-sharable object, which could move the object from a busy to an enable state.

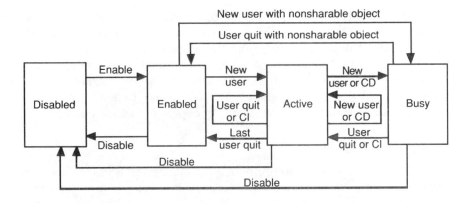

CI = capacity increase
CD = capacity decrease

Figure 11.5 Configuration management operation status states.

The disabled state occurs when a managed object is declared inoperable. For example, a network component might exhibit some defect that is beyond a permissible threshold and unacceptable to network control. A managed object can be declared disabled from any operational state. For example, it could even be declared disabled again because a second problem is noted. This could occur in many network components. One that comes to mind is a number of software bugs that manifest themselves as a program is diagnosed.

Figure 11.5 also shows the action of user quit, which allows the managed object to move from a busy to enabled state or from active to enable state. Just because a user quits does not mean the managed object need become disabled or inactive.

The notations CI and CD mean capacity increase and capacity decrease, respectively. These events can cause the managed object to move from active to busy state, from busy-to-active state, or remain in the busy state.

Configuration management also relies on X.731 to describe three administrative states (see Figure 11.6):

- Unlocked: The managed object can be used.
- Locked: The managed object cannot be used.
- Shutting down: The managed object can be used by current users but cannot be used by new users.

As the reader might expect, managed objects need not exhibit all three possible administrative states. That is, some resources may not be locked; therefore, they could not exhibit the locked state. As an example and as a general rule, a local area network read-only file server would not be allowed to exist in a locked state except for subsets of the file that might have security and access restrictions. In addition, certain resources may not be allowed to be shut down gracefully, and there would be no such thing as a shutting-down state. An example of this situation would be an administrative logical channel on a packet switch card that must exist in a perpetually unlocked state. The 0 channel on an X.25 interface comes to mind. Any graceful shutting down might create havoc on the dependent logical channels.

Security management

This facility is concerned with protecting the managed objects. It provides the rules for authentication procedures, the maintenance of access control routines, the supporting of the management of keys for encipherment, the maintenance of authorization facilities, and the maintenance of security logs. It is in the formative stages, but it is certain that it will rely extensively on the Directory service standards (X.500) for security support.

Performance management

This facility is more complete than some of the others. As suggested in the title, it supports the gathering of statistical data and applies the data to various analysis routines to measure the performance of the system.

It permits the use of models to determine (1) if a system is meeting the required throughput, (2) if a system is providing adequate response time, (3) if a system is approaching overload, and (4) if a system is being used efficiently.

The performance management facility relies on many definitions and concepts that have been developed for the other layers of OSI, such as residual error rate, transit delay, connection establishment delay, etc. Many of these definitions are widely used in X.25 and connectionless networks. In

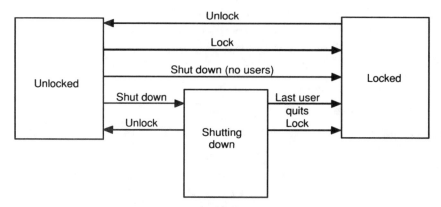

Figure 11.6 Configuration management administrative states.

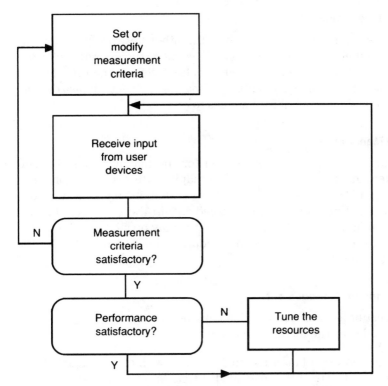

Figure 11.7 Performance management monitoring and tuning model.

addition, this standard provides directions on how to apply sampling formula to the analysis.

At the broadest level, performance management is organized around monitoring, analysis, and tuning functions. Figure 11.7 shows the organization of the performance management functions. The *workload monitoring* function will be described shortly.

The *throughput* monitoring function is used to measure the throughput on a communications circuit or a network node. It is defined for each direction of transfer as follows: The sending throughput is the ratio of successfully transferred protocol data units, within a sequence of data units provided at the maximum rate, to the time between the first and last request for the unit transfers. This assumes that all units measured are transmitted without errors.

Systems Management

The system management recommendations are descriptions of many of the important OSI management functions. Their organization is shown in Figure 11.1.

X.730 Object Management

X.730 lays the foundations for several of the other network management standards. It defines managed objects and describes the operations for the creation and configuration of managed objects. It defines the rules for the creation, deletion, renaming, and listing of managed objects. It also defines how to delete and make attribute changes to objects.

X.731 State Management

This standard identifies the state model for the managed objects. The OSI management standards permit two major state classifications: administrative and operational. Within these states an object can be defined as busy, active, enabled, or disabled within the operational state. Within the administrative state, the substates may be shutting down, unlocked, or locked. The X.731 Recommendation defines the rules for entering and leaving these states.

X.732 Relationship Management

This specification defines the relationships of the managed objects. The standard establishes the following relationships:

- Direct: Some portion of information associated with one managed object expressly identifies the other managed object.
- Indirect: A relationship is deduced between two managed objects by the concatenation of two or more direct relationships.
- Symmetric: Interaction rules between two managed objects are the same. For example, an interaction rule could stipulate the right to change an attribute of an object.
- Asymmetric: Interaction rules between two managed objects are different.

   In addition to these definitions, relationship management defines the service aspects of the management relationships with the create, get, and set services.

X.733 Alarm Reporting

This document defines and identifies five basic categories of errors (notification types), which comprise the X.733 kernel functional unit:

- Communications type
- Quality of service type
- Processing type

- Equipment type
- Environmental type

In addition to these five notification types, alarm reporting provides information on: (1) the probable cause of the alarm, (2) the severity of the alarm, (3) information on back-up status (if any), (4) information on threshold information (if any exists), (5) proposed repair actions, (6) indications if the alarm created state changes at the monitored object, and (7) textual information.

X.734 Event Reporting

This standard establishes the components to support remote event reporting and local event processing. The standard is formed around the concept of a set of *discriminators.*

As an example, the event forwarding discriminator is responsible for filtering events based on a number of selection criteria and deciding if the event is to be reported. In so doing, the discriminator uses a discriminator construct that establishes the thresholds and other criteria that must be satisfied for the event to be forwarded.

X.735 Log Control Function

As the title suggests X.735 defines the log control operations for a network management system. This specification defines how to preserve information about events and operations on managed objects. The standard also specifies mechanisms to control times for when logging occurs, for resuming logging, and for suspending logging. In addition it defines operations for retrieving and deleting log records, as well as the modification of criteria that are used in creating the logging records.

X.736 Security Alarm Reporting Function

This document defines procedures for the security management user to receive notifications on various types of alarms on managed objects. The standard specifies five types of security alarms:

- An *integrity* violation indicates that there has been a potential interruption in information flow. The information may have been deleted, inserted with other information, or modified in a way that is not permitted.

- An *operational* violation indicates that a requested service was not obtained because of a malfunction or otherwise incorrect invocation of a service.

- A *physical* violation indicates that a breach of a physical resource (for example, the tapping of a channel) has been detected.

- A *security service* violation indicates that some security device in the network has detected a security attack.

- The time to remain violation indicates that some event has occurred outside a permitted time threshold.

In addition to the security alarm types, this standard defines a number of causes that relate to the alarm types. Some examples follow:

- An indication that information was not expected but has been received

- An indication that information that is expected has not been received

- An indication that information that has been received has been modified illegally

- An indication that information has been received more than once, thereby alerting the user to a replay attack

- An indication that a communications medium has been tapped

X.740 Security Audit Trail Function

This specification is similar to the log control function, except that it provides audit trail logs that are a record of historical events that relate to the security measures. It provides support on auditing information related to accounting, security utilization, disconnections, connections, and other management operations.

X.741 Objects and Attributes for Access Control

The purpose of this document is to allow the network administrator to prevent unauthorized access to certain managed objects. It defines access control mechanisms based on a knowledge of a user's profile. In addition, this standard defines rules to allow or deny access to the managed objects.

X.742 Accounting Meter Function

This standard defines how charges, costs, and usage are identified and recorded in the managed network. In addition, it provides thresholds that limit the users' utilization of certain of the managed objects.

X.739 Workload Monitoring Function

This standard defines a workload monitoring model for a managed resource. It establishes procedures for feedback to the user in order to determine trends regarding the performance of the managed objects. It measures threshold values for reporting on early warning and severe warning conditions. It defines how to measure resource utilization and how to clear various conditions relating to the managed objects. It provides definitions for gauges, thresholds, counters, etc.

Structure of Management Information (SMI)

The SMI standards are divided into four parts. These parts are described in the following documents.

X.720 Management Information Model

SMI part 1 identifies the attributes of the managed objects that can be manipulated. It also identifies the operations of object attributes such as get, set, derive, add, and remove and the operations that may apply to the object itself, read, delete, and action.

For the newcomer to the X.700 Series, X.720 is one specification that is indispensable. In addition to the information cited in the previous paragraph, it also explains the basic concepts of this series and defines the key terms. As examples, inheritance, encapsulation, naming, naming containment, and the "packages" of attributes, operations, notifications, and behavior are defined in this Recommendation.

X.721 Definitions of Support Objects

This specification identifies common object classes used by OSI management. At the present time, it only defines the object class for discriminators: event, report, and service access.

X.723 Generic Management Information

This Recommendation is one of the more recent additions to the X.700 Series, having been published in November, 1993. Its purpose is to define managed objects that are common to more than one system; in effect managed object classes (MOCs). Included in X.723 are several ASN.1 templates or structures. As examples, the application process structure is coded to represent an element within a system; the connectionless-mode protocol machine structure is coded to represent a connectionless protocol operation, etc. Forty-six ASN.1 definitions are provided in this recommendation, so once again, the reader should have a knowledge of ASN.1, if you choose to study X.723.

X.722 Guidelines for the Definitions of Managed Objects

This document provides some very useful definitions of terms. It contains information on how to define objects, events, attributes, and actions. Additionally, it contains the rules on how to create and use templates.

X.710 Common Management Information Service Element (CMISE)

CMISE is defined in X.710. As the title suggests, it identifies the service elements used in management operations as well as their arguments (parameters). It also provides a framework for common management procedures that can be invoked from remote locations.

Introduction

The reader should remember that the OSI service element standards contain the rules for the creation and use of primitives between adjacent layers in the same machine. These primitives and their parameters are mapped into a PDU, which is transmitted across the communications link(s) to another machine.

CMISE is organized around the following types of services:

- M-EVENT-REPORT: This service is used to report an event to a service user. Because the operations of network entities are a function of the specifications of the managed objects, this event is not defined by the standard but can be any event about a managed object that the CMISE user chooses to report. The service provides the time of the occurrence of the event as well as the current time.

- M-GET: This service is used by CMISE to retrieve information from its peer. The service uses information about the managed object to obtain and return a set of attribute identifiers and values of the managed object or a selection of managed objects. It can only be used in a confirmed mode, and a reply is expected.

- M-CANCEL-GET: This service is invoked by the CMISE user to request a peer to cancel a previously requested M-GET service. It can only be used in a confirmed mode, and a reply is expected.

- M-SET: A CMISE user can use this service to request the modification of attribute values (the properties) of a managed object. It can be requested in a confirmed or nonconfirmed mode. If confirmed, a reply is expected.

- M-ACTION: This service is used by the user to request that another user perform some type of action on a managed object. It can be requested in a confirmed or nonconfirmed mode. If confirmed, a reply is expected.

- M-CREATE: This service is used to create a representation of another instance of a managed object, along with its associated management information values. It can only be used in a confirmed mode and a reply is expected.

- M-DELETE: This service performs the reverse operation of the M-CREATE. It deletes an instance of a managed object. It can only be used in a confirmed mode, and a reply is expected.

Use of underlaying services and layers

X.710 uses the following ROSE services: (1) RO-INVOKE, (2) RO-RESULT, (3) RO-ERROR, and (4) RO-REJECT. In turn, ROSE uses the P-DATA service of the presentation layer. Figure 11.8 shows the relationship of CMIP and CMISE to ACSE, ROSE, and the presentation layer.

Earlier versions of CMIP/CMISE defined services for the establishment of an application association. The services, shown in the shadowed box in Figure 11.8, are no longer defined. An association is obtained by the CMISE service user invoking the ACSE primitives directly.

X.711 Common Management Information Protocol (CMIP)

X.711 establishes the protocol specification for CMIP. CMIP supports the services listed in Table 11.2. These services were explained in the previous section. Notice that some of the services are confirmed, nonconfirmed, or have an option of using either the confirmed or nonconfirmed operation. These services allow two OSI management service users to set up actions to be performed on managed objects, to change attributes of the objects, and to report the status of the managed objects.

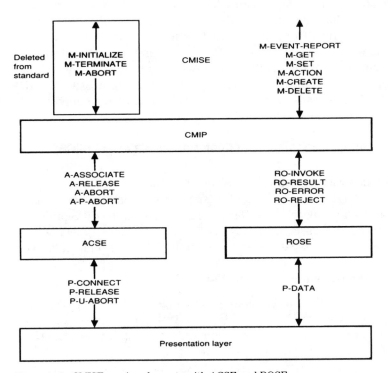

Figure 11.8 CMISE service elements with ACSE and ROSE.

TABLE 11.2 CMIP Services

Service	Type of confirmation
M-EVENT-REPORT	Confirmed/nonconfirmed
M-GET	Confirmed
M-SET	Confirmed/nonconfirmed
M-ACTION	Confirmed/nonconfirmed
M-CREATE	Confirmed
M-DELETE	Confirmed

Like the other OSI protocols, CMIP must follow rules on the composition and exchange of PDUs. All the CMIP PDUs are defined by ASN.1. The operations are defined in ROSE with the OPERATION and ERROR external macros. Because of its dependence on ROSE, CMIP does not contain state tables, event lists, predicates, or action tables.

Example of a CMIP PDU

Figure 11.9 is the ASN.1 coding for an m-Create PDU. The examination in this section will focus on an analysis of the line entries of the code.

The code shows three major entries. The *ARGUMENT CreateArgument* defines the arguments (parameters) that are carried in the data unit. These arguments are derived from the parameters in the CMISE *m-Create* request primitive. The *RESULT* and *ERRORS* contain information from the performing entity about the result of the create operation. These values are derived from the *m-Create* response primitive parameters issued by the performing CMISE service user. Further coding must be examined to understand how these values are used.

X.712 CMIP Protocol Implementation Conformance Statement (PICS) Proforma

This specification was released in the fall of 1992. As its name suggests, it defines the procedures for an organization (such as a test lab) to verify that a supplier's CMIP conforms to the X.711 Recommendation. The PICS proforma provides detailed checklists used to verify that the supplier supports all the CMIP services, and all the parameters cited in the X.711 Recommendation.

The Network Management Forum (NMF)

The NMF is a consortium of companies whose goal is to foster OSI standards. The forum was organized in July 1988, with the intent of accelerating the use of international computer communications standards. The NMF develops its standards based on the OSI Reference Model.

```
—Example of a CMIP Protocol Data Unit (simplified)
m-Create OPERATION
ARGUMENT        CreateArgument
RESULT          CreateResult
ERRORS          {accessDenied, classInstanceConflict,
                duplicateManagedObjectInstance,
                invalidAttributeValue, invalidObjectInstance
                missingAttributeValue, noSuchAttribute,
                noSuchObjectClass, noSuchObjectInstance,
                noSuchReferenceObject, processingFailure}
```

Figure 11.9 The M-CREATE PDU coded in ASN.1.

The membership of the NMF consists of many of the major and large telecommunications companies in North America, Japan, and Europe. Organizations such as British Telecom, AT&T, MCI, Nippon Telegraph and Telephone, Northern Telecom, DEC, and Telecom Canada are examples of the voting membership.

The NMF has published a series of documents describing the implementation of the OSI management standards. These documents are available from Network Management Forum at 40 Morristown Road, Bernardsville, NJ 07824 telephone 1 (201) 766-1544, FAX 1 (201) 766-5741. The organization has also developed a management information base (MIB) that defines many managed objects, such as communications links, transport layer class 4 connections, etc.

The NMF has published a seven-layer protocol suite, as well as numerous documents that define managed objects, protocols, functions, ASN.1 notations, rules for filtering and scoping, etc. Figure 11.10 shows the Forum suite. It is quite similar to other profiles that use the OSI layers.

Summary

The ITU-T network management standards are derived largely from the ISO network management standards. Both of these recommendations are published in one set of documents and are technically aligned with each other. The only exception to this statement is that the ITU-T does not publish a counterpart to the ISO's 10040 standard. The foundation for OSI network management standards are CMISE, published in X.710, and CMIP, published under X.711. The OSI network management standards make extensive use of object-oriented design techniques, which were summarized in this chapter.

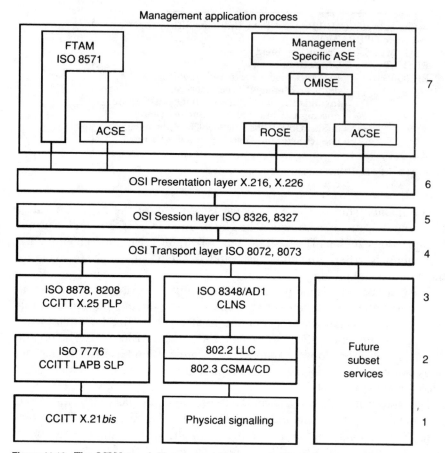

Figure 11.10 The OSI/Network Management Forum protocol stack.

12

The X.800 Series: OSI Applications

All the X.800 Recommendations have been published since 1991, although several of the specifications were available from the ISO in the late 1980s. The title of this series, "OSI Applications," also contains Recommendation X.800, which deals with the security architecture for these applications. This organization is reflected in Figure 12.1. The full titles of the X.800 Recommendations are listed below.

This chapter discusses these Recommendations as a logical whole and not on a one-by-one basis, because they are all closely interrelated.

- X.800: Security architecture for Open Systems Interconnection for CCITT applications

- X.851: Information technology—Open Systems Interconnection-Service definition for the commitment, concurrency and recovery service element

- X.852: Information technology—Open Systems Interconnection-Protocol for the commitment, concurrency and recovery service element: Protocol specification

- X.860: Open Systems Interconnection-Distributed transaction processing: Model

- X.862: Open Systems Interconnection-Distributed transaction processing: Protocol specification

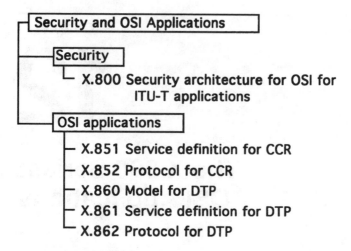

Where:
CCR is Commitment, concurrency and recovery
DTP is Distributed transaction processing

Figure 12.1 Organization of the X.800 Series.

X.800

X.800 provides the framework for building security features as part of the OSI Model. It describes which of these features are appropriate for each layer of the Model. The specification defines the specific *services* that are part of OSI security, and the specific *mechanisms* that are used to provide the services. Table 12.1 provides a summary of the services, and Table 12.2 provides a summary of the mechanisms. X.800 also provides illustrations and examples of the relationships of the security services and mechanisms. Lastly, this Recommendation defines several categories of security management, and how the services and mechanisms can be managed.

CCR

The CCR specifications describe the actions for managing activity between more than one application, perhaps across multiple sites. Most of the functions in CCR support the transfer and updating of elements (records, fields) in data bases. A key concept of CCR is the provision for *consistent states* for all applications and/or data involved in the activity. A consistent state means that the affected systems are accurate and correct and any duplicate copies contain the same values.

TABLE 12.1 X.800 Security Services

Peer entity authentication	Confirm the identify of a peer entity
Data origin authentication	Confirm that the claimed source of origin of data is valid
Access control service	Provide protection against unauthorized use of resources
Connection confidentiality	Provide confidentiality of all user data on a connection
Connectionless confidentiality	Provide confidentiality of a single service data unit (SDU)
Selective field confidentiality	Provide confidentiality of selected fields of data on a connection or within a single SDU of a connectionless operation
Traffic flow confidentiality	Provide confidentiality of data that can be observed in a traffic flow
Connection integrity with recovery	Provide integrity of all data on a connection, and detect and recover from any manipulation of the data
Connection integrity without recovery	Same as above, but no recovery is performed
Selective field connection integrity	Provide integrity of selected fields of an SDU on a connection
Connectionless integrity	Provide integrity of a single SDU
Selective field connectionless integrity	Provide integrity of selected fields within a single SDU
Non-repudiation, with proof of origin	Provide recipient with proof of the origin of data
Non-repudiation, with proof of delivery	Provide sender with proof of the delivery of data

TABLE 12.2 X.800 Security Mechanisms

Encypherment	Transformation of data to produce ciphertext (also called encryption)
Digital signature	Process of proving the source of data, and protection against forgery
Access control	Prevention of an unauthorized use of a resource
Data integrity	Preventing data from being altered or destroyed
Authentication exchange	Ensuring the identity of an entity
Traffic padding	Generation of spurious SDUs, connections, or spurious data within SDUs
Routing control	Choosing (perhaps avoiding) specific links or networks
Notarization	Using a third party to assure the proper delivery of data

In order to attain consistent states, CCR must be able to effect *concurrency control* and achieve *atomic actions*. Concurrency control is implemented by CCR through atomic action principles to ensure the affected system/datum does not have external changes made to it during an established state, and that all needed resources are committed to the full and complete execution of the state. It is irrelevant what the state is—it could be a data base update, multiple file transfers, the downloading of object code to packet switched, etc.

Another key concept of CCR assumes that an upper level protocol is available to assume the role of a master site (superior) in controlling the activity between the applications and/or data bases.

An example of the complexities and problems of data transfer, access, and update is provided here. Assume users A and B simultaneously access and update item XYZ in the data base. In the absence of control mechanisms, the data base reflects only the one update; the other update is lost. This happens when both users retrieve the data item, change it (add or subtract from the value), and write their revised value back into the data base.

The most common solution to this problem is to prevent sites A and B from simultaneous executions on the same data. Through the use of lockouts, for example, site B would not be allowed to execute until site A had completed its transaction.

Lockouts work reasonably well with a centralized data base. However, in a distributed environment, the sites may possibly lock each other out and prevent either transaction from completing its task. Mutual lockout is often called *deadly embrace*. Users A and B wish to update data base items Y and Z, respectively; consequently, user A locks data Y from user B and user B locks data Z from user A. To complete their transactions, both users need data from the other locked data bases. Hence, neither can execute further, and so the two sites are locked in a deadly embrace. Clearly, the deadly embrace is an unacceptable situation and the system must be able to detect, analyze, and resolve the problem. We address this problem shortly.

The two-phase commitment

The simple lockouts just described can be enhanced considerably by the concept of a two-phase commitment. CCR also uses a two-phase commitment operation in which:

- Phase 1: Superior determines which subordinates are to be involved in the activity and informs each subordinate of actions. Subordinate agrees or refuses to perform the actions.

- Phase 2: Superior orders the commitment or releases the resources to their beginning state.

CCR contains the concept of atomic action trees. With this approach, a subordinate can also assume the role of a superior to another subordinate, and each branch of the tree is an application association. CCR operates its atomic actions on the basis of a Boolean AND. All subordinates must agree in order for the atomic action to take place. If one or more refuses, the action is not committed and a rollback occurs.

If all subordinates commit, phase 2 is executed. During this phase, if any problems occur, a recovery process must be initiated. Be aware that the recovery operations are not defined in CCR but are application-specific.

DTP

The purpose of DTP is to support distributed transaction processing. It is built on the ability to detect failures in the processing of a transaction and to notify the DTP user about the nature of the failure (or success). The protocol has a number of authentication and security procedures, and permits multipart cooperation in supporting the transaction operations. It also allows the restoration of an object (such as a data base) to a prior, consistent state.

ACID

The DTP standard uses the concept of atomicity, consistency, isolation and durability known as ACID. The idea of ACID revolves around the concurrency, commitment, and recovery (CCR) transaction tree in which all parties must coordinate in the operation against the transaction and make a commitment to fully support it.

In the event of a problem, the object that has been operated upon must be restored to its original and consistent state. ACID also requires that a transaction be isolated from other transactions and they must not impact each other during the actual operation. Moreover, with ACID action must be finished before anything is revealed about it. Finally, ACID requires that all actions are performed successfully or none performed.

Functional units

DTP is organized around six functional units. The dialogue functional unit supports very basic DTP services, for example the initiation of dialogue, the sending of data, certain error routines, and certain abort operations.

The shared control functional unit permits both DTP service entities to issue primitives at any time. With polarized control, only one entity has control of the dialogue at any one time. This functional unit is based on the session layer control notion of having control of the token.

The handshake functional unit establishes rules for the DTP entities to synchronize and negotiate their operations with one another. The commit functional unit uses the CCR protocol for proper changing and rollbacking, if necessary.

Finally, the unchained transactions functional unit allows a DTP entity (designated as superior) to exclude a transaction subtree from certain types of transactions. Table 12.3 summarizes the major aspects of the DTP functional units.

TABLE 12.3 DTP Functional Units

- *Kernel*: The Dialogue functional unit supports the basic Transaction Services required to begin a dialogue, send data, signal a user- or provider-initiated error and end the dialog. User of provider abort may signal abnormal terminal.

- *Shared Control*: In the Shared Control functional unit both TPSUIs can issue request primitives subject only to the normal sequence constraints of the primitives. For example, data can be transferred by both TPSUIs at the same time.

- *Polarized Control*: In the Polarized Control functional unit, only one TPSUI has the control of the dialogue at any time. Many request primitives can be issued only by the TPSUI which controls the dialogue. This restriction is in addition to the normal sequence constraints for the primitives. For example, data can only be transferred by the TPSUI which has the control of the dialogue.

- *Handshake*: The Handshake functional unit allows a pair of TPSUIs to synchronzie their processing with one another.

- *Commit*: The Commit functional unit supports reliable commitment and rollback of provider supported transactions.

- *Unchained Transactions*: The Unchained Transactions functional unit allows a superior to exclude a transaction subtree from a sequence of provider supported transactions, and to re-include its direct subordinate in later provider supported transactions.

Summary

The X.800 Recommendations are quite similar to their earlier ISO counterparts. These principal operations are specified with X.800: (a) OSI security architecture; (b) commitment, concurrency and recovery; and (c) distributed transaction processing.

13

Conclusions

Ideally, an organization will ensure that a product conforms to a specification, standard, and/or recommendation before it acquires it. In reality, most organizations are not equipped to meet this goal. Therefore, some means must exist to test the product. Several organizations in Europe, the United States, and Japan now have the responsibility for the important job of conformance testing, that is, testing to see if the product performs in accordance with the international standard.

Introduction

On a general level, the reader should be aware that a data communications protocol that is "certified" by a testing organization does not mean it will operate with another seemingly "like" protocol. Conformance testing means that the product has been implemented in accordance with the standard and that it processes the relevant protocol data units (PDUs) as stated in the standard (see Figure 13.1). This means it operates satisfactorily with a testing machine.

Because of differences in other aspects in the implementation of products (timers, retries, etc.), two products may still not "interoperate." Consequently, some organizations also provide tests between two real systems to determine if they can communicate. This type of test is called *interoperability* testing (see Figure 13.2).

It should also be stressed that some tests consist of a *paper* evaluation, wherein the vendor's implementation conformance statement is evaluated against the standard. This approach works well enough if its limits are recognized, but the next step should be to test the product against the standard.

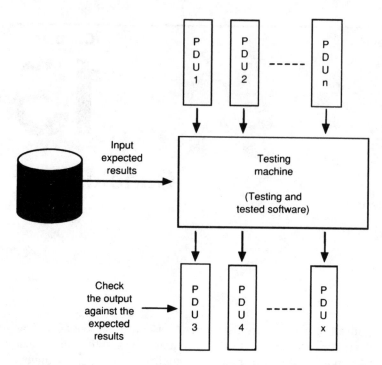

Figure 13.1 Conformance testing.

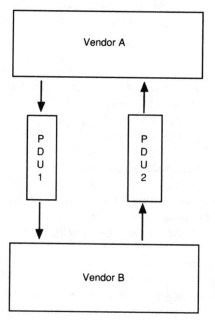

Figure 13.2 Interoperability testing.

Also, the reader should be aware that some forms of testing evaluate how the product performs in accepting, using, and creating the relevant PDUs. That is to say, the tests typically do not analyze the internal operations of the product. Therefore, the tests can reveal when something goes wrong, but they may not reveal why the problem occurred.

ITU-T X.290-294 Conformance Testing Methodology and Framework

The ITU-T recommendations and standards will be tested in accordance with X.290-X.294 Recommendations, which are listed below. (The MHS Recommendation X.480-X.485 define similar procedures.)

- X.290: OSI conformance testing methodology and framework for protocol Recommendations for ITU-T applications—General concepts

- X.291: OSI conformance testing methodology and framework for protocol Recommendations for ITU-T applications—Abstract test suite specification

- X.292: OSI conformance testing methodology and framework for protocol Recommendations for ITU-T applications—The tree and tabular combined notation (TTCN)

- X.293: OSI conformance testing methodology and framework for protocol Recommendations for ITU-T applications—Test realization

- X.294: OSI conformance testing methodology and framework for protocol Recommendations for ITU-T applications—requirements on test laboratories and clients for the conformance assessment process

The Protocol Implementation Conformance Statement

The PICS defines the capabilities and options that have been implemented in a system. It also describes any options that have been omitted. Its purpose is to provide a detailed description to allow the system to be tested. The PICS proformas are prepared as checklists or questionnaires, and are to be completed by the manufacturer or supplier of the OSI protocol that is under test. Annex A of X.291 provides guidance on how to prepare a PICS proforma.

Given that the PICS is prepared, a system under test (SUT) document is prepared. In accordance with X.290, the SUT describes four possible configurations for testing:

- Configuration 1: Uses OSI recommendations in all seven layers
- Configuration 2: Uses OSI recommendations in layers 1 to n

- Configuration 3: Uses OSI recommendations in layers 1 to 3 (for a network relay system) or in layers 1 to 7 (for an application relay system)
- Configuration 4: Combination of configurations 1 and 2 plus non-OSI protocols above layer n

X.290 also includes an implementation under test (IUT) document. This describes the specific parts of the system (layers and entities) that are to be tested. As the reader might expect, X.290 requires testing the operations of the PDUs and the abstract service primitives (ASPs) with the IUT.

The X.290 -X.294 Recommendations also provide the following information:

- As one would expect, a model for building a conformance testing system
- Establishment of test suites
- Specification of conformance clauses for the test suites
- Specifications for local and distributed testing
- Guidance on how to use testers
- Role of test laboratories

Summary

This book has provided a tutorial summary and reference guide to the ITU-T X Series Recommendations. It is hoped you have gained an understanding of these important standards and how they can be used in your data communications systems. As stated several times, this book should not serve as a substitute for the ITU-T documents. However, you should now be able to read the technical specification with more ease.

The ITU-T has done a laudable job in fostering and directing the development of these standards. Their use has led to decreased costs of communication interfaces as well as increased performance of data communications systems and networks.

Framework for Link Level Protocols

A prerequisite to any discussion of the data link control layers is an understanding of the high-level data link control (HDLC) specification. It forms the basis for several of ITU-T's link layer specifications such as link access protocol, balanced (LAPB), link access protocol on the D channel (LAPD), and link access protocol for modems (LAPM).

Introduction

HDLC is a line protocol specification published by the International Standards Organization (ISO) as ISO 3309 and ISO 4335 (and supporting documents 7809, 8471, 8885). It has achieved wide use throughout the world. The recommended standard provides for many functions and covers a wide range of applications. It is frequently used as a foundation for other protocols which use specific options in the HDLC repertoire.

This appendix addresses the main functions of HDLC. The reader is encouraged to check with specific vendors for their actual implementation of HDLC. Most vendors have a version of HDLC available, although the protocol is often renamed by the vendor or designated by different initials.

HDLC characteristics

HDLC provides for a number of link options to satisfy a wide variety of user requirements. It supports both half-duplex and full-duplex transmission, point-to-point and multipoint configuration, and switched and nonswitched channels.

An HDLC station is classified as one of three types:

- The *primary* station is in control of the data link. This station acts as a master and transmits *command* frames to the secondary stations on the channel. In turn, it receives *response* frames from those stations. If the link is multipoint, the primary station is responsible for maintaining a separate session with each station attached to the link.

- The *secondary* station acts as a slave to the primary station. It responds to the commands from the primary station in the form of responses. It maintains only one session, with the primary station, and has no responsibility for control on the link. Secondary stations cannot communicate directly with each other; they must first transfer their frames to the primary station.

- The *combined* station transmits both commands and responses and receives both commands and responses from another combined station. It maintains a session with the other combined station.

HDLC provides three methods to configure the channel for primary, secondary, and combined station use:

- An *unbalanced* configuration provides for one primary station and one or more secondary stations to operate as point-to-point or multipoint, half-duplex, full-duplex, switched or nonswitched. The configuration is called unbalanced because the primary station is responsible for controlling each secondary station and for establishing and maintaining the link.

- The *symmetrical* configuration is used very little today. The configuration provides for two independent, point-to-point, unbalanced station configurations. Each station has a primary and secondary status. Therefore, each station is considered logically to be two stations: a primary and a secondary station. The primary station transmits commands to the secondary station at the other end of the channel and vice versa. Even though the stations have both primary and secondary capabilities, the actual commands and responses are multiplexed onto one physical channel.

- A *balanced* configuration consists of two combined stations connected point-to-point only, half-duplex or full-duplex, switched or nonswitched. The combined stations have equal status on the channel and may send unsolicited frames to each other. Each station has equal responsibility for link control. Typically, a station uses a command in order to solicit a response from the other station. The other station can send its own command as well.

The terms *unbalanced* and *balanced* have nothing to do with the electrical characteristics of the circuit. In fact, data link controls should not be aware of the physical circuit attributes. The two terms are used in a completely different context at the physical and link levels.

While the HDLC stations are transferring data, they communicate in one of the three modes of operation:

- *Normal response mode* (NRM) requires the secondary station to receive explicit permission from the primary station before transmitting. After receiving permission, the secondary station initiates a response transmission which may contain data. The transmission may consist of one or more frames while the channel is being used by the secondary station. After the last frame transmission, the secondary station must again wait for explicit permission before it can transmit again.

- *Asynchronous response mode* (ARM) allows a secondary station to initiate transmission without receiving explicit permission from the primary station. The transmission may contain data frames or control information reflecting status changes of the secondary station. ARM can decrease overhead because the secondary station does not need a poll sequence in order to send data. A secondary station operating in ARM can transmit only when it detects an idle channel state for a two-way alternate (half-duplex) data flow or at any time for a two-way simultaneous (duplex) data flow. The primary station maintains responsibility for tasks such as error recovery, link setup, and link disconnections.

- *Asynchronous balanced mode* (ABM) uses combined stations. The combined station may initiate transmissions without receiving prior permission from the other combined station.

NRM is used frequently on multipoint lines. The primary station controls the link by issuing polls to the attached stations (usually terminals, personal computers, and cluster controllers). The ABM is a better choice on point-to-point links since it incurs no overhead and delay in polling. ARM is used very little today.

The term *asynchronous* has nothing to do with the format of the data and the physical interface of the stations. It is used to indicate that the stations need not receive a preliminary signal from another station before sending traffic. HDLC uses synchronous formats in its frames.

Frame format

HDLC uses the term *frame* to indicate the independent entity of data (protocol data unit, or PDU) transmitted across the link from one station to another. Figure A.1 shows the frame format. The frame consists of four or five fields:

Flag fields (F)	8 bits
Address field (A)	8 or 16 bits
Control field (C)	8 or 16 bits
Information field (I)	Variable length; not used in some frames
Frame check sequence field (FCS)	16 or 32 bits

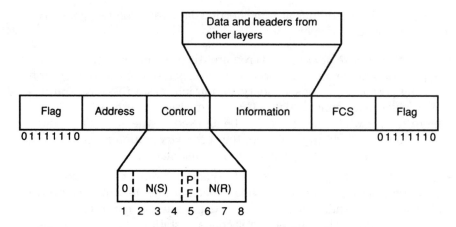

Flag: Delineates beginning and ending of frame
Address: Identifies link station
Control: Used for sequencing, flow control
Information: User data or control headers
FCS: Frame check sequence (for error checking)
N(S): Sending sequence number
N(R): Receiving sequence number
P/F: The poll or final bit

Figure A.1 The link layer frame.

All frames must start and end with the flag (F) sequence fields. The stations attached to the data link are required to continuously monitor the link for the flag sequence. The flag sequence consists of 01111110. Flags are continuously transmitted on the link between frames to keep the link in an active condition.

Other bit sequences are also used. At least seven, but less than fifteen, continuous 1s (abort signal) indicate a problem on the link. Fifteen or more 1s keep the channel in an idle condition. One use of the idle state is in support of a half-duplex session. A station can detect the idle pattern and reverse the direction of the transmission.

Once the receiving station detects a nonflag sequence, it is aware it has encountered the beginning of the frame, an abort condition, or an idle channel condition. Upon encountering the next flag sequence, the station recognizes it has found the full frame. In summary, the link recognizes the following bit sequences as:

01111110   = flags
At least 7, but less than 15 1s = abort
15 or more 1s   = idle

The time between the actual transmission of the frames on the channel is called interframe time fill. This time fill is accomplished by transmitting continuous flags between the frames. The flags may be 8-bit multiples and they can combine the ending 0 of the preceding flag with the starting 0 of the next flag.

HDLC is a code-transparent protocol. It does not rely on a specific code (ASCII/IA5, EBCDIC, etc.) for the interpretation of line control. For example, bit position n within an octet has a specific meaning, regardless of the other bits in the octet. On occasion, a flag-like field, 01111110, may be inserted into the user data stream (I field) by the application process. More frequently, the bit patterns in the other fields appear "flag-like." To prevent "phony" flags from being inserted into the frame, the transmitter inserts a 0 bit after it encounters five continuous 1s anywhere between the opening and closing flag of the frame. Consequently, zero insertion applies to the address, control, information, and FCS fields. This technique is called *bit stuffing*. As the frame is stuffed, it is transmitted across the link to the receiver.

The procedure to recover the frame of the receiver is a bit more involved (no pun intended). The "framing" receiver logic can be summarized as follows. The receiver continuously monitors the bit stream. After it receives a 0 bit with five continuous, succeeding 1 bits, it inspects the next bit. If it is a 0 bit, it pulls this bit out; in other words, it unstuffs the bit. However, if the seventh bit is a 1, the receiver inspects the eighth bit. If it is a 0, it recognizes that a flag sequence of 01111110 has been received. If it is a 1, it knows an abort or idle signal has been received and counts the number of succeeding 1 bits to take appropriate action.

In this manner, HDLC achieves code and data transparency. The protocol is not concerned about any particular bit code inside the data stream. Its main concern is to keep the flags unique.

Many systems use bit stuffing and the non-return-to-zero-inverted (NRZI) encoding technique to keep the receiver clock synchronized. With NRZI, binary 1s do not cause a line transition but binary 0s do. It might appear that a long sequence of 1s could present synchronization problems since the receiver clock would not receive the line transitions necessary for the clock adjustment. However, bit stuffing ensures that a 0 bit exists in the data stream at least every 5 bits. The receiver can use them for clock alignment.

The address (A) field identifies the primary or secondary station involved in the frame transmission or reception. A unique address is associated with each station. In an unbalanced configuration, the address field in both commands and responses contains the address of the secondary station. In balanced configurations, a command frame contains the destination station address and the response frame contains the sending station address.

The control (C) field contains the commands, responses, and sequence numbers used to maintain the data flow accountability of the link between the primary stations. The format and the contents of the control field vary, depending on the use of the HDLC frame.

The information (I) field contains the actual user data. The I field only resides in the frame under the Information frame format. It is usually not found in the supervisory or unnumbered frame.

One option of HDLC allows the I field to be used with an unnumbered information (UI) frame. This is a very important feature of HDLC, because it provides a capability to use the unnumbered frame to achieve a connectionless-mode operation at the link level. Several subsets of HDLC, such as logic link control (LLC), and LAPD, use the UI frame.

The FCS field is used to check for transmission errors between the two data link stations. The FCS field is created by a cyclic redundancy check. We summarize it here. The transmitting station performs modulo 2 division (based on an established polynomial) on the A, C, and I fields plus leading zeros and appends the remainder as the FCS field. In turn, the receiving station performs a division with the same polynomial on the A, C, I, and FCS fields. If the remainder equals a predetermined value, the chances are quite good that the transmission occurred without any errors. If the comparisons do not match, it indicates a probable transmission error, in which case the receiving station sends a negative acknowledgment, requiring a retransmission of the frame.

The control field

Let us return to a more detailed discussion of the control field (C) because it determines how HDLC controls the communications process. The control field defines the function of the frame and therefore invokes the logic to control the movement of the traffic between the receiving and sending stations. The field can be in one of three formats:

- The *Information* format frame is used to transmit end user data between the two devices. The information frame may also acknowledge the receipt of data from a transmitting station. It also can perform certain other functions such as a poll command.

- The *Supervisory* format frame performs control functions such as the acknowledgment of frames, the request for the retransmission of frames, and the request for the temporary suspension of the transmission frames. The actual usage of the supervisory frame is dependent on the operational mode of the link (normal response mode, asynchronous balanced mode, asynchronous response mode).

- The *Unnumbered* format is also used for control purposes. The frame is used to perform link initialization, link disconnection, and other link control functions. The frame uses 5 bit positions, which allows for up to 32 commands and 32 responses. The particular type of command and response depends on the HDLC class of procedure.

The actual format of the HDLC determines how the control field is coded and used. The simplest format is the Information transfer format. The N(S) (send sequence) number indicates the sequence number associated with a transmitted frame. The N(R) (receive sequence) number indicates the sequence number that is expected at the receiving site.

Piggybacking, flow control, and accounting for traffic

HDLC maintains accountability of the traffic and controls the flow of frames by the state variables and sequence numbers. The traffic at both the transmitting and receiving sites is controlled by state variables. The transmitting site maintains a send state variable V(S), which is the sequence number of the next frame to be transmitted. The receiving site maintains a receive state variable V(R), which contains the number that is expected to be in the sequence number of the next frame. The V(S) is incremented with each frame transmitted and placed in the send sequence field in the frame.

Upon receiving the frame, the receiving site checks the send sequence number with its V(R). If the CRC passes and if V(R) = N(S), it increments V(R) by 1, places the value in the sequence number field in a frame, and sends it to the original transmitting site to complete the accountability for the transmission.

If the V(R) does not match the sending sequence number in the frame (or the CRC does not pass), an error has occurred, and a reject or selective reject with a value in V(R) is sent to the original transmitting site. The V(R) value informs the transmitting DTE of the next frame that it is expected to send (i.e., the number of the frame to be retransmitted).

The poll/final bit

The fifth bit position in the control field is called the P/F, or poll/final bit. It is only recognized when set to 1 and is used by the primary and secondary stations to provide a dialogue with each other.

The primary station uses the P bit = 1 to solicit a status response from a secondary station. The P bit signifies a poll. The secondary station responds to a P bit with data or a status frame, and with the F bit = 1. The F bit can also signify end of transmission from the secondary station under NRM.

The P/F bit is called the P bit when used by the primary station and the F bit when used by the secondary station. Most versions of HDLC permit one P bit (awaiting an F bit response) to be outstanding at any time on the link. Consequently, a P set to 1 can be used as a checkpoint. That is, the P = 1 means "Respond to me, because I want to know your status." Checkpoints are quite important in all forms of automation. It is the way the machine clears up ambiguity and perhaps discards copies of previously transmitted frames. Under some versions of HDLC, the device may not proceed further until the F bit frame is received, but other versions of HDLC (such as LAPB) do not require the F bit frame to interrupt the full-duplex operations.

How does a station know if a received frame with the fifth bit = 1 is an F or P bit? After all, it is in the same bit position in all frames. HDLC provides an elegantly simple solution. The fifth bit is a P bit and the frame is a command if the address field contains the address of the receiving station; it is an F bit and the frame is a response if the address is that of the transmitting station.

This destination is quite important because a station may react quite differently to the two types of frames. For example, a command (address of receiver, P = 1) usually requires the station to send back specific types of frames.

The following is a summary of the addressing rules:

- A station places its own address in the address field when it transmits a response.

- A station places the address of the receiving station in the address field when it transmits a command.

HDLC commands and responses

Table A.1 shows the HDLC commands and responses. They are briefly summarized here.

The Receive Ready (RR) is used by the primary or secondary station to indicate that it is ready to receive an information frame and/or acknowledge previously received frames by using the N(R) field. The primary station may also use the Receive Ready command to poll a secondary station by setting the P bit to 1.

The Receive Not Ready (RNR) frame is used by the station to indicate a busy condition. This informs the transmitting station that the receiving station is unable to accept additional incoming data. The RNR frame may acknowledge previously transmitted frames by using the N(R) field. The busy condition can be cleared by sending the RR frame.

The Selective Reject (SREJ) is used by a station to request the retransmission of a single frame identified in the N(R) field. This field also performs inclusive acknowledgment; all information frames numbered up to N(R)1 are acknowledged. Once the SREJ has been transmitted, subsequent frames are accepted and held for the retransmitted frame. The SREJ condition is cleared upon receipt of an I frame with a N(S) equal to V(R).

An SREJ frame must be transmitted for each erroneous frame; each frame is treated as a separate error. Remember that only one SREJ frame can be outstanding at a time. Therefore, to send a second SREJ would contradict the first SREJ because all I frames with N(S) lower than N(R) of the second SREJ would be acknowledged.

The Reject (REJ) is used to request the retransmission of frames starting with the frame numbered in the N(R) field. Frames numbered N(R)1 are all acknowledged.

TABLE A.1 HDLC Control Field Format

Control field bit format	Encoding								Commands	Responses
	1	2	3	4	5	6	7	8		
Information	0	-	N(S)	-		-	N(R)	-	I	I
Supervisory	1	0	0	0	•	-	N(R)	-	RR	RR
	1	0	0	1	•	-	N(R)	-	REJ	REJ
	1	0	1	0	•	-	N(R)	-	RNR	RNR
	1	0	1	1	•	-	N(R)	-	SREJ	SREJ
Unnumbered	1	1	0	0	•	0	0	0	UI	UI
	1	1	0	0	•	0	0	1	SNRM	
	1	1	0	0	•	0	1	0	DISC	RD
	1	1	0	0	•	1	0	0	UP	
	1	1	0	0	•	1	1	0		UA
	1	1	0	1	•	0	0	0	NR0	NR0
	1	1	0	1	•	0	0	1	NR1	NR1
	1	1	0	1	•	0	1	0	NR2	NR2
	1	1	0	1	•	0	1	1	NR3	NR3
	1	1	1	0	•	0	0	0	SIM	RIM
	1	1	1	0	•	0	0	1		FRMR
	1	1	1	1	•	0	0	0	SARM	DM
	1	1	1	1	•	0	0	1	RSET	
	1	1	1	1	•	0	1	0	SARME	
	1	1	1	1	•	0	1	1	SNRME	
	1	1	1	1	•	1	0	0	SABM	
	1	1	1	1	•	1	0	1	XID	XID
	1	1	1	1	•	1	1	0	SABME	

I = Information; RR = Receive Ready; REJ = Reject; RNR = Receive Not Ready; SREJ = Selective Reject; UI = Unnumbered Information; SNRM = Set Normal Response Mode; DISC = Disconnect; RD = Request Disconnect; Up = Unnumbered Poll; RSET = Reset; XID = Exchange Identification; DM = Disconnect Mode; • = The P/F Bit; NR0 = Non-Reserved 0; NR1 = Non-Reserved 1; NR2 = Non-Reserved 2; NR3 = Non-Reserved 3; SIM = Set Initialization Mode; RIM = Request Initialization Mode; FRMR = Frame Reject; SARM = Set Async Response Mode; SARME = Set ARM Extended Mode; SNRM = Set Normal Response Mode; SNRME = Set NRM Extended Mode; SABM = Set Async Balance Mode; SABME = Set ABM Extended Mode.   UA = Unnumbered acknowledgment

The Unnumbered Information (UI) format allows for transmission of user data in an unnumbered (i.e., unsequenced) frame. The UI frame is actually a form of connectionless-mode link protocol in that the absence of the N(S) and N(R) fields preclude flow-controlling and acknowledging frames. The IEEE 802.2 LLC protocol uses this approach with its LLC type 1 version of HDLC.

The Request Initialization Mode (RIM) format is a request from a secondary station for initialization to a primary station. Once the secondary station sends RIM, it can monitor frames but can only respond to SIM, DISC, TEST, or XID.

The Set Normal Response Mode (SNRM) places the secondary station in the Normal Response Mode (NRM). The NRM precludes the secondary station from sending any unsolicited frames. This means the primary station controls all frame flow on the line.

The Disconnect (DISC) places the secondary station in the disconnected mode. This command is valuable for switched lines; the command provides a function similar to hanging up a telephone. UA is the expected response.

The Disconnect Mode (DM) is transmitted from a secondary station to indicate it is in the disconnect mode (not operational).

The Test (TEST) frame is used to solicit testing responses from the secondary station. HDLC does not stipulate how the TEST frames are to be used. An implementation can use the I field for diagnostic purposes, for example.

The Set Asynchronous Response Mode (SARM) allows a secondary station to transmit without a poll from the primary station. It places the secondary station in the information transfer state (IS) of ARM.

The Set Asynchronous Balanced Mode (SABM) sets the mode to ABM, in which stations are peers with each other. No polls are required to transmit since each station is a combined station.

The Set Normal Response Mode Extended (SNRME) sets SNRM with two octets in the control field. This is used for extended sequencing and permits the N(S) and N(R) to be 7 bits in length, thus increasing the window to a range of 1 to 127.

The Set Asynchronous Balanced Mode Extended (SABME) sets SABM with two octets in the control field for extended sequencing.

The Unnumbered Poll (UP) polls a station without regard to sequencing or acknowledgment. A response is optional if the poll bit is set to 0. This frame provides for one response opportunity.

The Reset (RESET) is used as follows: The transmitting station resets its N(S) and the receiving station resets its N(R). The command is used for recovery. Previously unacknowledged frames remain unacknowledged.

HDLC timers

The vendors vary in how they implement link level timers in a product. HDLC defines two timers, T1 and T2. Most implementations use T1 in some fashion. T2 is used but not as frequently as T1. The timers are used in the following manner:

T1A primary station issues a P bit and checks to see if a response to the P bit is received within a defined time. This function is controlled by the timer T1 and is called the "wait for F" time-out.

T2A station in the ARM mode that issues I frames checks to see whether acknowledgments are received within a timer period. This function is controlled by timer T2 and is called "wait for N(R)" time-out.

Since ARM is not used much today, timer T1 is typically invoked to handle the T2 functions.

HDLC Schema and HDLC "Subsets"

Many other link protocols are derived from HDLC. The practice has proved to be quite beneficial to the industry because it has provided a "baseline" link control standard from which to operate. In some companies, their existing HDLC software has been copied and modified to produce HDLC variations and subsets for special applications. However, be aware that while these link control systems are referred to as subsets, they sometimes include other capabilities not found in HDLC.

The major published subsets of HDLC are summarized in this section. The overall HDLC schema is shown in Figure A.2. Two options are provided for unbalanced links NRM (UN) and ARM (UA) and one for balanced ABM (BA).

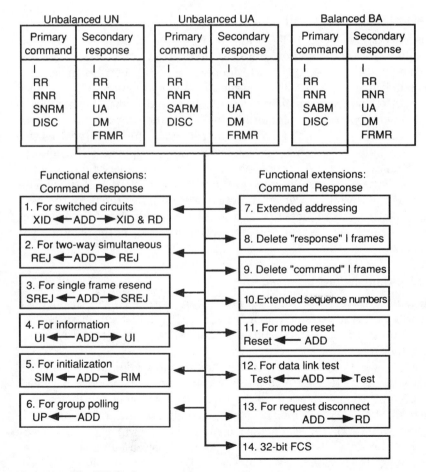

Figure A.2 The HDLC schema.

In order to classify a protocol conveniently, the terms UN, UA, and BA are used to denote which subset of HDLC is used. In addition, most subsets use the functional extensions. For example, a protocol classified as UN 3,7 uses the unbalanced normal response mode option and the selective reject and extended address functional extensions.

From an examination of Figure A.2, it is evident that HDLC provides a wide variety of options. Consequently, the full range of functions are not implemented as a single product. Rather, a vendor chooses the subset that best meets the need for the link protocol. The reader should be aware that using an HDLC product does not guarantee link compatibility with another vendor's HDLC product because each may implement a different subset of HDLC. Furthermore, some vendors implement features not found in the HDLC standards. When in doubt, read the manuals.

Link access procedure (LAP)

LAP is an earlier subset of HDLC that is based on the HDLC SARM command in an unbalanced configuration. It is classified UA 2,8 except it does not use the DM response. LAP is still used to support some X.25 network links.

To establish an LAP data link, the sending end (primary function) transmits an SARM in the control field to the receiving end (secondary function). Concurrent with the transmission of the SARM, the primary function will start a noresponse timer (T1). When the secondary function receives the SARM correctly, it transmits an acknowledgment response (UA: unnumbered acknowledgment). Receipt of the UA by the primary function confirms the initiation of one direction of the link and resets the T1 timer. The receipt of the SARM in a given direction will be interpreted by the secondary function as a request to initiate the other direction of transmission so the procedure may be repeated in the other direction at the discretion of the secondary function.

Link access procedure, balanced

LAPB is used by many private and public computer networks throughout the world. LAPB is classified as a BA 2,8 or BA 2,8,10 subset of HDLC. Option 2 provides for simultaneous rejection of frames in a two-way transmission mode.

Option 8 does not permit the transmitting of information in the response frames. This restriction presents no problem in an asynchronous balanced mode because the information can be transferred in command frames, and since both stations are combined stations, both can transmit commands. Moreover, with LAPB, the sending of a command frame with the P bit = 1 occurs when the station wants a "status" frame and not an information frame. Consequently, the responding station is not expected to return an I field.

LAPB is the link protocol layer for an X.25 network. It is used extensively worldwide and is found in many vendor's ports on a chip, with the X.25 network level software. An extensive discussion on LAPB can be found in *The X.25 Protocol* by Uyless Black, published by the IEEE Computer Society.

Logical link control (IEEE 802.2 and ISO 8802)

LLC is a standard sponsored by the IEEE 802 standards committee for local area networks. The standard permits the interfacing of a local area network to other local networks as well as to a wide area network. LLC uses a subclass of the HDLC superset. LLC is classified as BA-2,4.

LLC permits three types of implementations of HDLC: type 1, using the UI frame (unacknowledged connectionless service); type 2, using the conventional I frame (acknowledged connection-oriented service); and type 3, using AC frames (acknowledged connectionless service).

LLC is intended to operate over a peer-to-peer multipoint channel using the UI or SABME frames. Therefore, each frame contains the address of both the sending and receiving station.

Link access procedure, D channel

LAPD is another subset of the HDLC structure, although it has extensions beyond HDLC. It is derived from LAPB. LAPD is used as a data link control for the integrated services digital network (ISDN).

ISDN provides LAPD to allow DTEs to communicate with each other across the ISDN D channel. (Many users want LAPD for B channels as well.) It is specifically designed for the link across the ISDN user-network interface.

LAPD has a very similar frame format to HDLC and LAPB. Moreover, it provides for unnumbered, supervisory, and information transfer frames. LAPD also allows a modulo 128 operation. The control octet to distinguish between the Information, Supervisory, and Unnumbered formats is identical to HDLC.

	Commands	Responses
Type 1	UI	
	XID	XID
	TEST	TEST
Type 2 (I format)	I	I
(S format)	RR	RR
	RNR	RNR
	REJ	REJ
(U format)	SABME	
	DISC	UA
		DM
		FRMR
Type 3	AC	AC

LLC frame types.

LAPD provides for two octets for the address field. This is valuable for multiplexing multiple functions onto the D channel. Each ISDN basic access can support up to eight stations. The address field is used to identify the specific terminal and service access point (SAP). The address field contains the address field extension bits, a command/response indication bit, a service access point identifier (SAPI), and a terminal end-point identifier (TEI). These entities are discussed in the following paragraphs.

The purpose of the address field extension is to provide more bits for an address. The presence of a 1 in the first bit of an address field octet signals that it is the final octet of the address field. Consequently, a two-octet address would have a field address extension value of 0 in the first octet and a 1 in the second octet. The address field extension bit allows the user of both the SAPI in the first octet and the TEI in the second octet.

The command/response (C/R) field bit identifies the frame as either a command or a response. The user side sends commands with the C/R bit set to 0. It responds with the C/R bit set to 1. The network does the opposite—it sends commands with C/R set to 1 and responses with C/R set to 0.

The SAPI identifies the point where the data link layer services are provided to the layer above (that is, layer 3). (If your understanding of the concept of the SAPI is vague, review chapter 1.)

The TEI identifies a specific connection within the SAP. It can identify either a single terminal (TE) or multiple terminals. The TEI is assigned by a separate assignment procedure. Collectively, the TEI and SAPI are called the data link connection identifier (DLCI), which identifies each data link connection on the D channel. As stated earlier, the control field identifies the type of frame as well as the sequence numbers used to maintain windows and acknowledgments between the sending and receiving devices.

Presently, the SAPI and TEI values are allocated as follows:

SAPI	
Value	Related entity
0	Call Control Procedures
16	Packet Procedures
32–47	Reserved for National Use
63	Management Procedures
Others	Reserved

TEI	
Value	User type
0–63	Non-Automatic Assignment
64–126	Automatic Assignment

Two commands and responses in LAPB do not exist in the HDLC schema. These are sequenced information 0 (SI0) and sequenced information 1 (SI1). The purpose of the SI0/SI1 commands is to transfer information using sequentially acknowledged frames. These frames contain information fields provided by layer 3. The information commands are verified by the means of the end (SI) field. The P bit is set to 1 for all SI0/SI1 commands. The SI0 and SI1 responses are used during single frame operation to acknowledge the receipt of SI0 and SI1 command frames and to report the loss of frames or any synchronization problems. LAPD does not allow information fields to be placed in the SI0 and SI1 response frames. Obviously, information fields are in the SI0 and SI1 command frames.

LAPD differs from LAPB in a number of ways. The most fundamental difference is that LAPB is intended for point-to-point operating user DTE-to-packet exchange (DCE). LAPD is designed for multiple access on the link. The other major differences are summarized as follows:

- LAPB and LAPD use different timers.
- As explained earlier, the addressing structure differs.
- LAPD implements the HDLC unnumbered information frame (UI).
- LAPB uses only the sequenced information frames.
- LAPD primitives

LAPD uses a number of primitives for its communications with the network layer, the physical layer, and a management entity which resides outside both layers. The primitives are summarized in Table A.2.

TABLE A.2 LAPD Primitives

Primitive	Function
DL-ESTABLISH (level 2/3 boundary)	Issued on the establishment of frame operations
DL-RELEASE (level 2/3 boundary	Issued on the termination of frame operations
DL-DATA (boundary)	Used to pass data between layers (level 2/3 with acknowledgments)
DL-UNIT-DATA (level 2/3 boundary)	Used to pass data with no acknowledgments
MDL-ASSIGN (level management/2 boundary)	Used to associate TEI value with a specified end point
MDL-REMOVE (level management/2 boundary)	Removes the MDL-ASSIGN
MDL-ERROR (level management/2 boundary)	Associated with an error that cannot be corrected by LAPD
MDL-UNIT-DATA (level management/2 boundary)	Used to pass data with no acknowledgments
PH-DATA (level 2/1 boundary)	Used to pass frames across layers
DH-ACTIVATE (level 2/1 boundary)	Used to set up physical link
PH-DEACTIVATE (level 2/1 boundary)	Used to deactivate physical link

LAPM with V.42

V.42 implements a link control protocol called LAPM. It is based on the HDLC family of protocols and was also written from the LAPB specification, which is part of X.25.

The principal difference between LAPM and a conventional HDLC implementation relates to the use of the address field. The address field consists of the data link identifier, the C/R bit, and the address extension bit. The C/R bit is a command/response bit which identifies the frame as either a command or response. The DLCI value is used to transfer information between the X.24 interfaces. Currently, DLCI is set to 0 to identify a DTE-to-DTE interface. Other values are permitted within the limits defined in the recommendation. The EA bit can be set to 1 to designate another octet for DLCI.

B

Tutorial on ISDN

The reader should remember that ITU-T intends the integrated services digital network (ISDN) to complement its X Series Recommendations for data services. The complementary functions include user facilities, quality of service features, and call progress signals such as those established in X.2 and X.96. This is not to say that the existing X Series features cannot be enhanced. For example, with the multiple terminal arrangement established through ISDN basic access and 64-kbit/s data rate, it is certainly reasonable to expect that the basic X services could be enhanced by ISDN features.

Although less explicitly stated, it will be evident in several parts of this book that ITU-T intends to make the transition from analog to digital services as transparent as possible to the end user. For example, the terminal adapter (TA) is an integral part of many of the V and X Series Recommendations. We will see its function is to provide a transition interface from the current analog interfaces to ISDN.

The ISDN terminal

To begin the analysis of ISDN, consider the end user ISDN terminal in Figure B.1. This device (called data terminal equipment, or DTE, in this book) is identified by the ISDN term *TE1* (terminal equipment, type 1). The TE1 connects to the ISDN through a twisted pair four-wire digital link. This link uses time division multiplexing (TDM) to provide three channels, designated as the B, B, and D channels (or 2B + D). The B channels operate at a speed of 64 kbit/s; the D channel operates at 16 kbit/s. The 2B + D is designated as the basic rate interface. ISDN also allows up to eight TE1s to share one 2B + D channel.

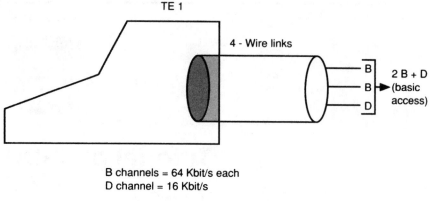

B channels = 64 Kbit/s each
D channel = 16 Kbit/s

2B + D = 144 Kbit/s as the basic rate

TE1 = terminal equipment 1 (an ISDN device)

Figure B.1 ISDN basic access.

Figure B.2 illustrates the format for the basic access D channel. The information (I) field carries upper layer information. With an ISDN connection, it carries the ISDN network layer. It could also carry an X.25 packet.

Figure B.3 illustrates other ISDN options. In this scenario, the user DTE is a TE2 device, which is currently in use in such equipment as IBM 3270 terminals, telex devices, etc. The TE2 connects to the TA, which allows non-ISDN terminals to operate over ISDN lines. The user side of the TA typically uses a conventional physical layer interface, such as EIA-232, or a V Series specification. It is packaged like an external modem or as a board that plugs into an expansion slot on the TE2 devices. The EIA or V Series interface is called the R interface in ISDN terminology.

Basic access and primary access

The TA and TE2 devices are connected through the basic access to either an ISDN network termination 1 (NT1) or NT2 device. The NT1 is a customer premise device which connects the four-wire subscriber wiring to the conventional two-wire local loop. ISDN allows up to eight terminal devices to be addressed by NT1.

The NT1 is responsible for the physical layer functions (of OSI), such as signalling synchronization and timing. It provides a user with a standardized interface.

The NT2 is a more intelligent piece of customer premise equipment. It is typically found in a digital PBX and contains the layers 2 and 3 protocol functions. The NT2 device is capable of performing concentration services.

It multiplexes 23 B + D channels onto the line at a combined rate of 1.544 Mbit/s. This function is called the ISDN primary rate access.

The NT1 and NT2 devices may be combined into a single device called NT12. This device handles the physical, data link, and network layer functions.

In summary, the TE equipment is responsible for user communications and the NT equipment is responsible for network communications.

ISDN reference points and interfaces

The reference points are logical interfaces between the functional groupings. The S reference point is the 2 B + D interface into the NT1 or NT2 device. The T interface is the reference point on the customer side of the NT1 device. It is the ISDN "plug in the wall." It is the same as the S interface on the basic rate access lines. The U interface is the reference point for the two-wire side of the NT1 equipment. It separates an NT1 from the line termination (LT) equipment. The V reference point separates the line termination (LT) from the exchange termination (ET) equipment.

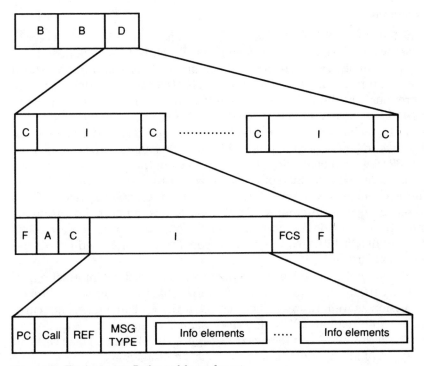

Figure B.2 The basic rate D channel frame format.

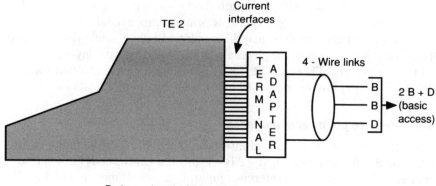

Figure B.3   The ISDN terminal adapter.

ISDN channels

The most common ISDN interface supports a bit rate of 144 kbit/s. The rate includes two 64-kbit/s B channels, and one 16-kbit/s D channel. In addition to these channels, ISDN provides for framing control and other overhead bits, which totals to a 192-kbit/s bit rate. The 144-kbit/s interfaces operate synchronously in the full-duplex mode over the same physical connector. The 144-kbit/s signal provides time division multiplexed provisions for the two 64-kbit/s channels and one 16-kbit/s channel. The standard allows the B channels to be further multiplexed in the subchannels. For example, 8-, 16-, or 32-kbit/s subchannels can be derived from the B channels. The two B channels can be combined or broken down as the user desires.

The B channels are intended to carry user information streams. They provide for several different kinds of applications support. For example, channel B can provide for voice at 64 kbit/s, data transmission for packet-switch utilities at bit rates less than or equal to 64 kbit/s, and broadband voice at 64 kbit/s or less.

The D channel is intended to carry control and signalling information, although in certain cases, ISDN allows for the D channel to support user data transmission as well. However, be aware that the B channel does not carry signalling information. ISDN describes signalling information as s-type, packet data as p-type, and telemetry as t-type. The D channel may carry all these types of information through statistical multiplexing.

ISDN layers

The ISDN approach is to provide an end user with full support through the seven layers of the OSI Model. In so doing, ISDN is divided into two kinds of services—the bearer services, responsible for providing support for the lower three levels of the seven-layer standard, and teleservices (for example, telephone, teletex, Videotex message handling), responsible for providing support through all seven layers of the model and generally making use of the underlying lower-level capabilities of bearer services. The services are referred to as low-layer and high-layer functions, respectively. The ISDN functions are allocated according to the layering principles of the OSI and ITU-T standards.

It is hoped this brief tutorial on the major ISDN terms and concepts will be sufficient for the reader to cope with these ideas as they relate to the X Series Recommendations.

C

Overview of the V Series Recommendations

The V Series Recommendations are titled: *Data Communication over the Telephone Network*. The title describes their functions: to define conventions and procedures for the transfer of data using the public telephone network. Because the vast majority of telephone systems use analog signalling to support voice communications, a substantial part of the V Series Recommendations is devoted to defining the conventions for converting digital data signals (binary digits, or bits) into analog signals, and vice versa. The ITU-T publishes the V Series in Fascicle VII.1.

Table C.1 lists the broad categories of functions and services described in the V Series. The recommendations are organized into six sections. Each of these sections is further divided into the specific recommendation. This approach facilitates the organization and use of the documents.

The V Series Recommendations are identified by a title (sometimes rather lengthy) and a two- or three-digit number, which is preceded by the letter V. The V numbers are associated with the six categories:

General:	V.1–V.7
Interfaces and Voice-band Modems:	V.10–V.33
Wide-band Modems:	V.35–V.37
Error Control:	V.40–V.42
Transmission Quality and Maintenance:	V.50–V.57
Interworking with other Networks:	V.100–V.230

TABLE C.1 Organization of the V Series

Section number and name	Description
1. General	General overview descriptions of coding, symbols, signalling rates, power levels used over the telephone network
2. Interfaces & Voice-band Modems	Detailed descriptions of the interfaces between modems, and their DTEs, as well as the signalling conventions between the modems over voice-band frequencies
3. Wideband Modems	Detailed descriptions of the interfaces between wideband modems, and their DTEs, as well as the signalling conventions between modems
4. Error Control	Descriptions of conventions to obtain error detection and error correcting services
5. Transmission Quality	Specifications on testing methods, noise and measurement, maintenance limits on telephone circuits
6. Interworking with Other Networks	Specifications for interworking telephone networks with packet and ISDN networks

Principal Functions of the V Series Recommendations

Many of the V Series Recommendations describe the following physical layer functions:

- Description of the procedures for data transfer across the interface between the DTE and the DCE (in ITU-T terms, the *interchange circuits*)
- Provision for *control* interchange circuits between the devices to govern how the *data* interchange circuits operate
- Description of clocking signals on specific interchange circuits to synchronize data flow and regulate the bit transfer rate between the DTE and the DCE
- Description of signals to synchronize data flow and regulate the bit transfer rate between the two DCE.
- Description of electrical ground
- Description of the mechanical connectors (such as pins, sockets, and plugs)

The ITU-T V Series physical layer protocols describe four attributes of the interface: electrical, functional, mechanical, and procedural. The *electrical* attributes describe the voltage (or current) levels, the timing of the electrical signals, and all other electrical characteristics (capacitance, signal rise time, etc.).

The *functional* attributes describe the functions to be performed by the interchange circuits at the physical interface. Many physical layer protocols classify these functions as control, timing, data, and ground. The *mechanical* attributes describe the dimensions of the connectors and the number of wires on the interface. Usually, the data, signalling, and control wires are enclosed in one cover. The *procedural* attributes describe what the connectors must do and the sequence of events required to effect the data transfer across the interface.

The V Series Recommendations are concerned with analog transmissions schemes and, therefore, devote considerable amount of space to modem operations. A modem is responsible for providing the translation and interface between the digital and analog worlds. The term *modem* is derived from (1) the process of accepting digital bits from the local DTE and changing them into a form suitable for analog transmission (modulation) and (2) receiving the signal at the other modem and transforming it back to its original digital representation (demodulation) for the remote DTE. The term modem is derived from the two words *mod*ulator and *dem*odulator.

Several of the new V Series are concerned with interworking the V Series modems with an ISDN. These recommendations are similar to some of the X Series Interfaces Recommendations and are found in V.100 through V.230.

A companion book by the same author is devoted to the V Series Recommendations: *The V Series Recommendations*, McGraw-Hill, 1991.

D

ITU-T X Series Recommendations

This appendix lists all the ITU-T X Series Recommendations. These documents can be obtained from:

ITU-T Sales Department
Place des Nations
CH-1211 Geneve 20 Suisse
Telephone: 730.52.85
FAX: 730.51.94

Other organizations (commercial services) offer these documents, but the ITU-T prices are lower than the prices of these organizations. Some of the documents are available in some electronic libraries, but be aware that the material is copyrighted.

X.1–X.39: Services, facilities, and interfaces

X.1: International user classes of service in, and categories of access to, public data networks and integrated services digital networks (ISDNs)

X.2: International data transmission services and optional user facilities in public data networks and ISDNs

X.3: Packet assembly/disassembly (PAD) facility in a public data network

X.4: General structure of signals of International Alphabet No. 5 code for character-oriented data transmission over public data networks

X.5: Facsimile packet assembly/disassembly facility (FPAD) in a public data network

X.6: Multicast service definition

X.7: Technical characteristics of data transmission services

X.10: Categories of access for data terminal equipment (DTE) to public data transmission services

X.20: Interface between data terminal equipment (DTE) and data circuit-terminating equipment (DCE) for start-stop transmission services on public data networks

X.20 bis: Use on public data networks of data terminal equipment (DTE) which is designed for interfacing to asynchronous duplex V-Series modems

X.21: Interface between data terminal equipment (DTE) and data circuit-terminating equipment (DCE) for synchronous operation on public data networks

X.21 bis: Use on public data networks of data terminal equipment (DTE) which is designed for interfacing to synchronous duplex V-Series modems

X.22: Multiplex DTE/DCE interface for user classes 36

X.24: List of definitions for interchange circuits between data terminal equipment (DTE) and data circuit-terminating equipment (DCE) on public data networks

X.25: Interface between data terminal equipment (DTE) and data circuit-terminating equipment (DCE) for terminals operating in the packet mode and connected to public data networks by dedicated circuit

X.26: Electrical characteristics for unbalanced double-current interchange circuits for general use with integrated circuit equipment in the field of data communications

X.27: Electrical characteristics for balanced double-current interchange circuits for general use with integrated circuit equipment in the field of data communications

X.28: DTE/DCE interface for a start/stop mode data terminal equipment accessing the packet assembly/disassembly facility (PAD) in a public data network situated in the same country

X.29: Procedures for the exchange of control information and user data between a packet assembly/disassembly (PAD) facility and a packet mode DTE or another PAD

X.30: Support of X.21 and X.21 bis based data terminal equipments (DTEs) by an integrated services digital network (ISDN)

X.31: Support of packet mode terminal equipment by an ISDN

X.32: Interface between data terminal equipment (DTE) and data circuit-terminating equipment (DCE) for terminals operating in the packet mode and accessing a packet switched public data network through a public switched telephone network or an integrated services digital network or a circuit switched public data network

X.35: Interface between a PSPDN and a private PSDN which is based on X.25 procedures and enhancements to define a gateway function that is provided in the PSPDN

X.38: G3 facsimile equipment/DCE interface for G3 facsimile equipment accessing the facsimile packet assembly/disassembly facility (FPAD) in a public data network situated in the same country

X.39: Procedures for the exchange of control information and user data between a facsimile packet assembly/disassembly (FPAD) facility and a packet mode data terminal equipment (DTE) or another FPAD

X.40: Standardization of frequency-shift modulated transmission systems for the provision of telegraph and data channels by frequency division of a group (NLIF)

X.50–X.181: Transmission, signalling and switching; network aspects; maintenance; and administrative arrangements

X.50: Fundamental parameters of a multiplexing scheme for the international interface between synchronous data networks

X.50 bis: Fundamental parameters of a 48 kbit/s user data signalling rate transmission scheme for the international interface between synchronous data networks

X.51: Fundamental parameters of a multiplexing scheme for the international interface between synchronous data networks using 10-bit envelope structure

X.51 bis: Fundamental parameters of a 48 kbit/s user data signalling rate transmission scheme for the international interface between synchronous data networks using 10-bit envelope structure

X.52: Method of encoding anisochronous signals into a synchronous user bearer

X.53: Numbering of channels on international multiplex links at 64 kbit/s

X.54: Allocation of channels on international multiplex links at 64 kbit/s

X.55: Interface between synchronous data networks using a 6 + 2 envelope structure and single channel per carrier (SCPC) satellite channels

X.56: Interface between synchronous data networks using an 8 + 2 envelope structure and single channel per carrier (SCPC) satellite channels

X.57: Method of transmitting a single lower speed data channel on a 64 kbit/s data stream

X.58: Fundamental parameters of a multiplexing scheme for the international interface between synchronous nonswitched data networks using no envelope structure

X.60: Common channel signalling for circuit switched data applications

X.61: Signalling system No. 7-Data user part

X.70: Terminal and transit control signalling system for start-stop services on international circuits between anisochronous data networks

X.71: Decentralized terminal and transit control signalling system on international circuits between synchronous data networks

X.75: Packet-switched signalling system between public networks providing data transmission services

X.80: Interworking of interexchange signalling systems for circuit switched data services

X.81: Interworking between an ISDN circuit switched and a circuit switched public data network (CSPDN)

X.82: Detailed arrangements for interworking between CSPDNs and PSPDNs based on Recommendation T.70

X.92: Hypothetical reference connections for public synchronous data networks

X.96: Call progress signals in public data networks

X.110: International routing principles and routing plan for public data networks

X.121: International numbering plan for public data networks

X.122: Numbering plan for the E.164 and X.121 numbering plans

X.130: Call processing delays in public data networks when providing international synchronous circuit-switched data services

X.131: Call blocking in public data networks when providing international synchronous circuit-switched data services

X.134: Portion boundaries and packet layer reference events for defining packet-switched performance parameters

X.135: Speed of service (delay and throughput) performance values for public data networks when providing international packet-switched services

X.136: Accuracy and dependability performance values for public data networks when providing international packet-switched services

X.137: Availability performance values for public data networks when providing international packet-switched services

X.138: Measurement of performance values for public data networks when providing international packet-switched services

X.139: Echo, drop, generator and test DTEs for measurement of performance values in public data networks when providing international packet switched services

X.140: General quality of service parameters for communication via public data networks

X.141: General principles for the detection and correction of errors in public data networks

X.150: Principles of maintenance testing for public data networks using data terminal equipment (DTE) and data circuit-terminating equipment (DCE) test loops

X.180: Administrative arrangements for international closed user groups (CUGs)

X.181: Administrative arrangements for the provision of international permanent virtual circuits (PVCs)

X.200 – X.294: Open Systems Interconnection (OSI)—General

X.200: Reference Model of Open Systems Interconnection for ITU-T Applications

X.207: Information technology—Open Systems Interconnection—Application layer structure

X.208: Specification of Abstract Syntax Notation One (ASN.1)

X.209: Specification of basic encoding rules for Abstract Syntax Notation One (ASN.1)

X.210: Information technology—Open Systems Interconnection—basic reference model—conventions for the definition of OSI services

X.211: Physical service definition of Open Systems Interconnection for ITU-T applications

X.212: Data link service definition for Open Systems Interconnection for ITU-T

X.213: Information technology—network service definition for Open Systems Interconnection

X.214: Information technology—Open Systems Interconnection—transport service definition

X.215: Session service definition for Open Systems Interconnection for ITU-T applications

X.216: Presentation service definition for Open Systems Interconnection for ITU-T applications

X.217: Service definition for the association control service element

X.218: Reliable transfer: Model and service definition

X.219: Remote operations: Model, notation and service definition

X.220: Use of X.200-Series protocols in ITU-T applications

X.223: Use of X.25 to provide the OSI connection-mode network service for ITU-T applications

X.224: Protocol for providing the OSI connection-mode transport service

X.225: Session protocol specification for Open Systems Interconnection for CCITT applications

X.226: Presentation protocol specification for Open Systems Interconnection for CCITT applications

X.227: Connection-oriented protocol specification for the association control service element

X.228: Reliable transfer: protocol specification

X.229: Remote operations: protocol specification

X.233: Information technology—Protocol for providing the connectionless-mode network service: protocol specification

X.237: Connectionless protocol specification for the association control service element

X.244: Procedure for the exchange of protocol identification during virtual call establishment on packet switched public data networks

X.248: Reliable transfer service element—Protocol implementation conformance statement (PICS) proforma

X.249: Remote operations service element—Protocol implementation conformance statement (PICS) proforma

X.264: Transport protocol identification mechanism

X.283: Elements of management information related to the OSI network layer

X.290: OSI conformance testing methodology and framework for protocol Recommendations for ITU-T applications—General concepts

X.291: OSI conformance testing methodology and framework for protocol Recommendations for ITU-T applications—Abstract test suite specification

X.292: OSI conformance testing methodology and framework for protocol Recommendations for ITU-T applications—The tree and tabular combined notation (TTCN)

X.293: OSI conformance testing methodology and framework for protocol Recommendations for ITU-T applications—Test realization

X.294: OSI conformance testing methodology and framework for protocol Recommendations for ITU-T applications-requirements on test laboratories and clients for the conformance assessment process

X.300–X.370: Interworking

X.300: General principles for interworking between public networks, and between public networks and other networks for the provision of data transmission services

X.301: Description of the general arrangements for call control within a subnetwork and between subnetworks for the provision of data transmission services

X.302: Description of the general arrangements for internal network utilities within a subnetwork and intermediate utilities between subnetworks for the provision of data transmission services

X.305: Functionalities of subnetworks relating to the support of the OSI connection-mode network service

X.320: General arrangements for interworking between integrated services digital networks (ISDNs) for the provision of data transmission services

X.321: General arrangements for interworking between circuit switched public data networks (CSPDNs) and integrated service digital networks (ISDNs) for the provision of data transmission services

X.322: General arrangements for interworking between packet switched public data networks (PSPDNs) and circuit switched public data networks (CSPDNs) for the provision of data transmission services

X.323: General arrangements for interworking between packet switched public data networks (PSPDNs)

X.324: General arrangements for interworking between packet switched public data networks (PSPDNs) and public mobile systems for the provision of data transmission services

X.325: General arrangements for interworking between packet switched public data networks (PSPDNs) and integrated services digital networks (ISDNs) for the provision of data transmission services

X.326: General arrangements for interworking between packet switched public data networks (PSPDNs) and common channel signalling network (CCSN)

X.327: General arrangements for interworking between packet switched public data networks (PSPDNs) and private data networks for the provision of data transmission services

X.340: General arrangements for interworking between a packet switched public data network (PSPDN) and the international telex network

X.350: General interworking requirements to be met for data transmission in international public mobile satellite systems

X.351: Special requirements to be met for packet assembly/disassembly facilities (PADs) located at or in association with coast earth stations in the public mobile satellite service

X.352: Interworking between packet switched public data networks and public maritime mobile satellite data transmission systems

X.353: Routing principles for interconnecting public maritime mobile satellite data transmission systems with public data networks

X.370: Arrangements for the transfer of internetwork management information

X.400–X.485: Message handling systems

X.400: Message handling systems and service overview

X.402: Message handling systems: Overall architecture

X.403: Message handling systems: Conformance testing

X.407: Message handling systems: Abstract service definition conventions

X.408: Message handling systems: Encoded information type conversion rules

X.411: Message handling systems: Message transfer system: Abstract service definition and procedures

X.413: Message handling systems: Message store: Abstract-service definition

X.419: Message handling systems: Protocol specifications

X.420: Message handling systems: Interpersonal messaging system

X.435: Message handling systems: Electronic data interchange messaging system

X.440: Message handling systems: Voice messaging system

X.480: Message handling systems and directory services—Conformance testing

X.481: P2 protocol: Protocol implementation conformance statement (PICS) proforma

X.482: P1 protocol: Protocol implementation conformance statement (PICS) proforma

X.483: P3 Protocol: Protocol implementation conformance statement (PICS) proforma

X.484: P7 Protocol: Protocol implementation conformance statement (PICS) proforma

X.485: Message handling systems: Voice messaging system protocol implementation conformance statement (PICS) proforma

X.500–X.582: Directory services

X.500: The Directory—Overview of concepts, models, and services

X.501: The Directory—Models

X.509: The Directory—Authentication framework

X.511: The Directory—Abstract service definition

X.518: The Directory—Procedures for distributed operation

X.519: The Directory—Protocol specifications

X.520: The Directory—Selected attribute types

X.521: The Directory—Selected object classes

X.525: The Directory—Replication

X.581: Directory access protocol—Protocol implementation conformance statement (PICS)

X.582: Directory system protocol—Protocol implementation conformance statement (PICS)

X.610–X.665: OSI-networking and system aspects

X.610: Provision and support of the OSI connection-mode network service

X.612: Information technology—Provision of the OSI connection-mode network service by packet-mode terminal equipment connected to an integrated services digital network (ISDN)

X.613: Information technology—Use of X.25 packet layer protocol in conjunction with X.21/X.21 bis to provide the OSI connection-mode network service

X.614: Information technology—Use of X.25 packet layer protocol to provide the OSI connection network service over the telephone network

X.650: Open Systems Interconnection (OSI)-Reference model for naming and addressing

X.660: Information technology—Open Systems Interconnection-Procedures for the operation of OSI Registration Authorities-General procedures

X.665: Information technology—Open Systems Interconnection-Procedures for the operation of OSI Registration Authorities: Application processes and application entities

X.700–X.745: OSI management

X.700: Management framework for Open Systems Interconnection (OSI) for CCITT applications

X.701: Information technology—Open Systems Interconnection-Systems management overview

X.710: Common management information service definition for CCITT applications

X.711: Common management information protocol specification for CCITT applications

X.712: Information technology—Open Systems Interconnection-Common management information protocol: Protocol implementation conformance statement proforma

X.720: Information technology—Open Systems Interconnection-Structure of Management Information: Management information model

X.721: Information technology—Open Systems Interconnection-Structure of Management Information: Definition of management information

X.722: Information technology—Open Systems Interconnection-Structure of Management Information: Guidelines for the definition of managed objects

X.723: Information technology—Open Systems Interconnection-Structure of Management Information: Generic management information

X.724: Information technology—Open Systems Interconnection-Structure of Management Information-Requirements and guidelines for implementation conformance statement proformas associated with OSI management

X.730: Information technology—Open Systems Interconnection-Systems Management: Object management function

X.731: Information technology—Open Systems Interconnection-Systems Management: State management function

X.732: Information technology—Open Systems Interconnection-Systems Management: Attributes for representing relationships

X.733: Information technology—Open Systems Interconnection-Systems Management: Alarm reporting function

X.734: Information technology—Open Systems Interconnection-Systems Management: Event report management function

X.735: Information technology—Open Systems Interconnection-Systems Management: Log control function

X.736: Information technology—Open Systems Interconnection-Systems Management: Security alarm reporting function

X.738: Information technology—Open Systems Interconnection-Systems Management: Summarization function

X.739: Information technology—Open Systems Interconnection-Systems Management: Metric objects and attributes

X.740: Information technology—Open Systems Interconnection-Systems Management: Security audit trail function

X.745: Information technology—Open Systems Interconnection-Systems Management: Test management function

X.800–X.862: Security and OSI applications

X.800: Security architecture for—Open Systems Interconnection for CCITT applications

X.851: Information technology—Open Systems Interconnection-Service definition for the commitment, concurrency and recovery service element

X.852: Information technology—Open Systems Interconnection-Protocol for the commitment, concurrency and recovery service element: Protocol specification

X.860: Open Systems Interconnection—Distributed transaction processing: Model

X.862: Open Systems Interconnection—Distributed transaction processing: Protocol specification

Index

A

AbandonFailed, 236-237
access unit (AU), 189
accounting management, 263-264
accounting meter function, 270
additional physical rendition service, 208
addressing, 39, 178
  message handling system, 197
administration operation, 201, 205
agent process, 258-259
alarm reporting, 268-269
alternate recipient allowed service, 211
alternate recipient assignment service, 211
application entity (AE), 192
application layer, 8, 17, 124, 149-155
  aspects of, 150-155
  data integrity issues, 158
application service element (ASE), 192, 259, 262
applications, OSI, 277-282
Asian-Oceanic Workshop (AOW), 18
ASN.1, 143-147, 242
association control service element (ACSE), 150-152, 193-195, 239, 259, 262
connectionless, 152
asynchronous transfer mode (ATM), 9
atomicity, consistency, isolation, and durability (ACID), 281
AttributeError, 236
attributes, 197
  access control, 270
  managed objects, 257
authentication procedures, directory access, 239-241
authorizing users indication service, 211
auto-forwarded indication service, 211
automatic request for repeat, 116

B

basic physical rendition service element, 212
bilateral closed user group (BCUG), 35
blind copy recipient indication service element, 212
Blue Books, major changes in 1988, 19-21
body part encryption indication service element, 212

C

call deflection subscription, 35
call redirection, 36
calling line identification, 35
cancel deferred delivery operation, 200, 205
chained service operations, 237
change credentials operation, 202, 205
charging information, 36
circuit switched public data network (CSPDN), 173, 178
closed user group (CUG), 36
commitment, concurrency, and recovery (CCR), 278-281
common channel signalling network (CCSN), 173, 183
common management information protocol (CMIP), 262, 273-274
common management information service element (CMISE), 262, 272
communications, managing computer, 227-232
compare operation, 234
configuration management, 264-266
confirm, 124
conformance testing, 18, 121, 285-286
connectionless-mode operations, 8-9
connectionless network protocol (CLNP), 18
connection-oriented operations, 8-9

ABOUT THE AUTHOR

Uyless Black is the founder of the Information Engineering Institute in Virginia. He is the author of numerous books and articles on computer communications, and lectures and consults worldwide on the subject. He is McGraw-Hill's series advisor for the McGraw-Hill Series on Computer Communications.